# Essential
# Geographical
# Skills

**Darren Christian**

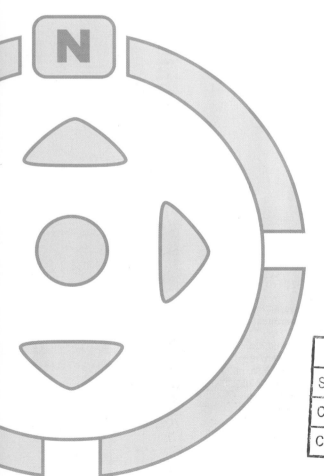

Nelson Thornes

First published in 2009 by:
Nelson Thornes Ltd
Delta Place
27 Bath Road
CHELTENHAM
GL53 7TH
United Kingdom

09  10  11  12  13  /  10  9  8  7  6  5  4  3  2  1

A catalogue record for this book is available from the British Library

ISBN 978 1 4085 0333 1

Cover photograph/illustration by Getty Images copyright Ron Chapple
Illustrations by Oxford Designers & Illustrators and Tech-Set Limited
Page make-up by Tech-Set Limited

Printed and bound in Croatia by Zrinski

## Acknowledgements

Special thanks to Joceline Bury, Jon Taylor and Frank Barclay. Their help and support has been invaluable. Thanks also to Sarah Chapman and Geoff Covey for their contributions.

Darren Christian

### Photograph acknowledgements

**Blom Oblique Imagery:** p15a; p15b; p122b; **Corbis:** EDAW consortium/Handout/Reuters/ p112b; NASA / p121b; **Crown Copyright material is reproduced with the permission of the Controller, Office of Public Sector Information (OPSI):** p128a;
**Darren Christian:** p16a; p16b; p17a; p22a; p28a; p33a; p119a; p120a; p121a; p130b;
**Fotolia:** p34a; p118a; p118b; p118c; **Getty Images:** p112a; National Geographic/p133a;
**Google Earth (earth.google.uk):** p37b; **iStockphoto:** p11b; p67a; p67b; p67c; p77a; p82a; p106a; p134a; p136a; **Joceline Bury:** p43b; **Paul Cornish (www.thegeographer.co.uk):** p25b; **Pete Carr:** p11a; **Pulse Creative Limited:** p60a; **Science Photo Library:** Daniel Sambraus/p98a; NASA/p132a; Simon Fraser/p92a; University of Dundee/p122a.

**Advisory Unit resources:** p123a; p125a; p126a; p129a. Reproduced by permission of the Advisory Unit. AEGIS 3 GIS software is developed and published by The Advisory Unit: Computers in Education, The Innovation Centre, Hatfield, AL10 9AB.

**Ordnance Survey maps:** p37c; p38b; p40a; p40b; p41a; p45a; p55a; p56b; p146a; p147a; p148a; p149a; Reproduced by permission of Ordnance Survey on behalf of HMSO. © Crown copyright 2009. All rights reserved. Ordnance Survey Licence number 100017284. Experian Goad Digital Plans include mapping data licensed from Ordnance Survey with the permission of the Controller of Her Majesty's Stationery Office. © Crown copyright and Experian Copyright. All rights reserved. Licence number PU 100017316, or such other copyright statement notified from time to time to the Client by Experian.

# Contents

# Foreword

There has been a significant shift in education towards developing and improving students' acquisition of a range of skills and competences. At the same time in geography, there has been a shift away from the submission of coursework as part of the assessment package for GCSE and A-level. At A-level, this has been replaced by formal examinations that test students' ability to understand and apply a variety of geographical skills to both familiar and unfamiliar contexts. At GCSE, there are still a variety of geographical skills to acquire, but there is also the relatively new concept of the controlled assessment.

This book is designed to incorporate the new subject-specific skills at GCSE and A-level. It is an essential guide to the geographical skills and techniques students need to show evidence of acquiring, at both GCSE and AS/A2.

All skills and techniques in this book are distilled down to their simplest form in order to make them as accessible as possible to all students. The book is written for and directed at the students. However, this does not change the fact that some techniques are quite difficult to understand and apply. The key to success is in following the sequence that leads to the completion of each skill.

When working through this book, a simple rule is to stop as soon as something does not make sense. Students are then advised to re-read the section and if things are still unclear, ask for help. By practising and carefully applying each of the skills they need to learn, students will be well prepared for their examinations.

Darren Christian

# How to use this book

*Essential Geographical Skills* is a practical course companion designed to accompany you through four years of geographical study, all the way from GCSE to A-level. The book shows you where and how to apply the full range of geographical skills that you will need to know and be able to demonstrate for your exams. Case studies with worked examples will guide you step by step through the principles of using each skill. Useful tips and hints throughout will help you carry out the tasks set, which will reinforce your learning of each skill and technique.

Whatever stage you are at with your geographical studies, you are guaranteed to find this book useful. The best way to use the book really depends on what you are preparing for. For example, if you are preparing for the skills-based examinations at A-level, you should focus on the relevant sections of chapter one and also the specific skills your specification requires. Do this by making a detailed list and start by practising the skills which you feel you know least about. Alternatively, simply use the index to find the specific skill you are looking for and then practise working with it.

Each section of the book includes the following features:

- **In this section you will learn**
  This is a checklist of what the section covers.

- **Difficulty symbol**
  The beginning of each section has one of the following symbols showing the level of difficulty of each skill. However, this is only a rough guide as different students face different challenges with the learning of new skills.

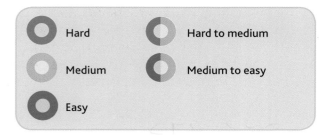

- Hard
- Hard to medium
- Medium
- Medium to easy
- Easy

- **Key terms**
  These list the technical terms used in the section, along with their definitions.

- **Key points**
  This feature gives useful information about the strengths and weaknesses of the skill, possible alternatives and important points to watch.

- **Tasks**
  A series of tasks or activities are included in each section. These are designed to give you practice in applying the skill you have learned to a range of data and situations. Once you have learned the skill, you can apply the methods to your own data or situation.

- **Tips**
  This feature offers advice or hints, for example on how you might use a technique in other contexts, or where a similar skill covered in another section might be used instead.

- **Formula**
  Where a technique requires a formula, the formula is given, along with notes on how to make sense of it.

- **Take it further**
  This gives you more information or ideas for extension work.

- **Summary**
  This gives a useful list of what the section has covered, and what you need to know in order to be able to apply the particular skill. Strengths and weaknesses are also considered where appropriate.

## Chapter 1
### Investigative skills

This chapter is concerned with geography fieldwork enquiry, at both GCSE and A-level. You can no longer submit a piece of coursework for assessment: at GCSE, all parts of your fieldwork have to be written up in school, sometimes under examination conditions; and at A-level you have to answer questions about your geographical enquiry in an examination.

It has therefore never been more important to make sure that you fully understand what you have done in your fieldwork and why you did it. This chapter outlines the stages in the enquiry process and shows you how to structure your enquiry write-up.

At A-level, while you may still write up your fieldwork enquiry in the traditional way, you will not receive a mark or grade that counts towards your overall AS or A2 grade. The key to success at AS and A2 is in having one eye on the sorts of questions you might be asked in the examination. Your teacher might not even ask you to write up the enquiry from start to finish. Instead you might well be guided towards using the enquiry experience to answer practice examination-style questions, building up a portfolio of materials designed to maximise your preparation for the examination. Whatever approach you take, it is vital to realise that your knowledge and understanding will be tested under formal examination-type conditions.

## Chapter 2
### Cartography: making the most of maps

This is about getting back to basics with maps and showing you many different ways in which they can be used effectively to display information and explore patterns. Maps are a simplified representation of the world around us, and to use them effectively and to use the techniques in this chapter effectively, you must pay attention to detail. You should also practise the skills and techniques as much as possible, in order to become a successful geographer.

As with the other chapters in this book, you should identify the skills you need for your particular piece of learning and work through them at your own pace. Some of the map skills overlap with each other, and these have therefore been grouped together. However you need only focus on the skills that are important for your specification, examination or piece of fieldwork.

## Chapter 3
### Using graphs to transform data

Creating and using a variety of graphs to show information is a really interesting aspect of geography. It enables you to transform a series or set of numbers into a visual representation of your data. Data usually makes little sense in a table unless there is only a very small amount of it. Once the data are displayed in a graphical form, however, it becomes possible to see patterns, identify trends and identify anomalies.

Attention to detail is essential. This chapter will help you avoid making simple mistakes such as mixing up your axes or mis-plotting information, which would not only make a mess of your data but would also be likely to cost you easy marks in an examination. Worse, some exam questions go on to ask you to describe the pattern you have just plotted. This can lead to a cumulative error whereby one minor mistake can lead to a series of other avoidable errors.

## Chapter 4
### Statistical skills

Statistical tests are an important additional tool available to the geographer. You can use them to analyse data (almost always numbers) generated by your study. They are particularly useful when only a small sample of data is available, helping you to infer relationships, and compare and contrast data sets. They can help you draw conclusions from your data sample and also to generate further research ideas. Once you have completed your analysis, you will be able to interpret your findings with greater certainty. The analysis will also help you determine what further work you need to do.

This chapter does not explain the mathematical principles underpinning these techniques, but aims to show how you – as a geographer – might use these techniques to analyse data. You should then be able to apply the techniques to your own data and, perhaps more importantly, use these techniques in external public examinations.

## Chapter 5
## Using information communication technology

New technologies have become invaluable tools for geographers. Geography teachers have been highly innovative in their use of ICT and the internet, which have become vital resources for both teaching and learning. There are literally thousands of websites available to the geographer for researching case studies, exploring concepts and ideas and revising for examinations. This section contains tips and hints on how to maximise your time when using the internet.

Geographical Information Systems (GIS) are a relatively new concept involving a range of resources such as satellite images, remote sensing and other modern technologies, and using them to locate, capture and manage geographical information. This section will help you understand the scope of GIS, and introduces some basic concepts and ideas about how to use GIS to present data as well as interpret patterns.

Your specifications expect you to be able to use programmes such as Microsoft Word, Excel and PowerPoint to organise and present your work. You also need to be able to use databases. The Census, Meteorological Office and Environment Agency websites are good examples of databases you should have experience of using during your course. These databases contain vast amounts of information. You will be shown how to search for specific information and display your findings.

## Chapter 6
## Preparing for success

This section looks at various aspects of revision and examination technique. The examination is the culmination of your work. It is the time when you have the opportunity to show an examiner what you have learned. The examiner knows nothing about you or the work you have put into your studies. All the examiner has to go on is the work that you produce under pressure in a timed examination.

At GCSE, your result might influence whether or not you can go on to do A-level or it might influence a college or an employer to give you a place. At A-level the stakes are even higher as you may well need a particular grade to get into university. Whatever your path, getting it right in the examination is crucial, and exam skills are as important as all the other geographical skills you need to learn.

## Appendices

A list of useful websites, advice on using OS maps, critical values tables and pages of graph paper are provided here.

## GCSE Controlled assessment

Controlled assessment is a new type of assessment that replaces the submission of GCSE coursework. It is worth 25 per cent of your GCSE.

You still have to do a fieldwork enquiry, but you write up the analysis and evaluation of your enquiry under controlled conditions, similar to exam conditions. Provided you stay well organised and follow the rules, tasks for controlled assessment can be an enjoyable and rewarding part of your GCSE course.

## What does controlled assessment look like?

Since different examination boards have slightly different approaches to the controlled assessment, you should consult your own specification for specific detail. The approach used here is for a single geographical enquiry, which includes the collection of primary data through fieldwork.

### The rules of controlled assessment

All examination boards have a responsibility to ensure that their own controlled assessment system meets the Qualifications and Curriculum Authority (QCA) rules. This system has three types of 'control': limited, medium and high. Control refers to the amount of freedom the examination board allows you and your teachers in the fieldwork set-up, subsequent write-up and marking. The control types are as follows:

#### Limited control

■ Your teacher will decide on the location of the study and the methodologies you will carry out.
■ You may work in groups while collecting data, but must write up the enquiry on your own.
■ There will be help available for you in researching and planning.

#### Medium control

■ Your teacher will mark your completed controlled assessment. You will find copies of the marking criteria in the specification you are using.
■ A sample is then sent off to the examination board for checking (moderation).

#### High control

■ The examination board sets the broad theme you have to investigate.
■ The analysis and evaluation part of the write-up has to be completed under highly controlled conditions (similar to formal examination conditions).

### Steps in controlled assessment

Whatever title for the controlled assessment your teacher obtains from the examination board, the steps you need to take will be broadly the same. The table opposite shows the typical structure of a GCSE controlled assessment from start to finish.

### In this section you will learn:

1 what controlled assessment is
2 how to write up a controlled assessment using a piece of local fieldwork
3 how the controlled assessment will be marked
4 the hints and tips for success.

### Key terms

**Quantitative data**: information that can be measured in numbers. For example, temperature measured in degrees Celsius is quantitative data.

**Qualitative data**: information related to thoughts and feelings. It is much more difficult to measure. For example, a questionnaire exploring people's perceptions and opinions would be qualitative in nature.

### The controlled assessment

■ has strict time limits
■ is led by your teacher
■ has sections you complete alone
■ changes each year
■ is around 2,000 words
■ involves fieldwork
■ requires independent research
■ is completed mainly in school time.

### Tip

Check the word length in your specification. This example assumes your controlled assessment will be around 2,000 words.

# The eight stages of controlled assessment

| Stage | Involves | Word count guide | Level of control |
|---|---|---|---|
| 1 The aim | A statement that outlines the overall purpose of the investigation. | N/A | The examination board will determine the specific area of your specification (high control), but your teacher will have some flexibility over the aim itself (limited control). |
| 2 The background to the study | Two parts: you will investigate the background to the study area; you will also outline the underlying theory, concept or idea that formed the basis of your study. | 200–400 words | You will be able to decide specific theories, concepts and ideas that you want to include (limited control). You will be free to contextualise your study in your own local area (limited control). |
| 3 Hypotheses/ objectives/ research questions | You can either: use statements that you will later accept or reject (hypotheses); or break down your aim into specific tasks to be carried out during the process (objectives); or ask research questions that you plan to answer subsequently. | N/A | Your hypotheses, objectives or research questions will be derived from the original aim, so there is lots of flexibility here (limited control). |
| 4 Methodologies | Collecting primary data and, where appropriate, secondary data (see page 13). You will use a combination of quantitative and qualitative data (see page 9). | 400–600 words | Methodologies are designed to address the hypotheses, objectives and research questions. There is lots of flexibility here, but these must be written up individually without help and during school time (limited control). |
| 5 Results | Displaying data in the form of tables of results. Results are transformed into graphs, maps and other representations of the data. | N/A | Creating your tables, and drawing graphs and maps can be done in school with lots of flexibility as to how you choose to display your findings. While you may share your primary and secondary data, you cannot work in groups on the graphs and maps themselves. You may use a computer to display findings, but check your own specification to make sure (limited control). |
| 6 Analysis and discussion | Analysis involves summarising and manipulating the data from your results. Discussion involves considering the patterns and trends in your data as well as exploring the anomalies thrown up by your results. | 600–800 words | This has to be written under formal supervision (high control). You will have access to the rest of your study, but no access to other secondary sources such as the internet (high control). |
| 7 Conclusions | In revisiting the original aim and hypotheses/ objectives/research questions, you have to assess the extent to which the findings match your expectations and whether you have achieved your overall aim. | 200 words | As above |
| 8 Evaluation | This can occur throughout the study or in a designated section at the end. You have to consider the strengths and weaknesses of the study design and methodologies. How far have the weaknesses in these areas affected the outcomes of the study and what improvements would you make if you did this study again? | 200–400 words | As above |

# A sample controlled assessment investigation

Using an example of a simple and broad title based on a human study, the following plan of a controlled assessment from start to finish shows you what is involved, how to get organised and how to score good marks. You can then take the ideas from this and apply them to your own study.

*Case study*

## How do different communities in Liverpool view the impact of the 2008 Capital of Culture on the city and its people?

The students were interested to see if their area of residence affected people's perceptions about recent changes in the city. They decided to base their study around a series of research questions:

1  Where are the more and less affluent parts of the city?
2  Are local people aware of the developments in Liverpool brought about by the 2008 Capital of Culture award?
3  How do residents in different parts of the city view these changes?
4  Are people in less affluent areas more likely to be negative about the impact of the Capital of Culture?
5  Does the number of venues and events visited affect perceptions about the impact of the Capital of Culture?

From these research questions, the students began to work on the background to their study. Students were advised to select at least three research questions and no more than five. They were also free to devise their own research questions.

### The background to the study

First, they collected these recent images of Liverpool to set the scene for their study.

Next, they obtained a more detailed map showing the study area where they collected their primary data. They were then able to annotate their map showing some of the characteristics of their study area. They used **www.multimap.com** to obtain their digital map.

Two images of Liverpool, European Capital of Culture 2008

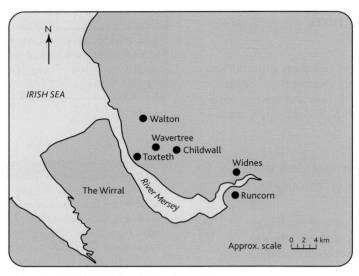

A simple sketch map identifying the study area

### Describing recent changes

The students performed an internet search to identify recent changes affecting Liverpool's population, mainly those brought about by the 2008 Capital of Culture. They started their search with the following phrase:

'Liverpool 2008 Capital of Culture.'

Working in groups, they soon began to focus on recent developments both in the city centre and around the city generally.

As the final part of the background they looked at recent media reports, charting changes in the local community and how these have affected the local people.

They now had some background knowledge of their study area and had completed the first three stages of their investigation.

### Tasks

**Collecting background information**

1  Write the aim for your enquiry.
2  Describe the characteristics of your study area. Include the following:
   ■ aspects of the local geography
   ■ recent changes to the area (if appropriate)
   ■ the underlying theory, concept or idea that you are investigating.
3  Suggest 3–4 of one of the following:
   ■ hypotheses
   ■ objectives
   ■ research questions.
   (For more information on this, see page 24.)
4  Obtain some images and maps of your study area.
5  Carefully annotate the resources showing important local features and factors that affected the choice of location for the study at different scales.

# Primary data collection: the field trip

Getting ready for your field trip is an exciting time. You will probably be given a checklist by your teacher, which tells you what you need to bring on the day. The type of field trip you undertake will determine the type of equipment you will need. Your school will almost certainly provide any specialist equipment. Be warned that forgetting even the smallest item (such as wellington boots for a river study) can mean that you might not be able to attend. You must take preparation seriously and always be prompt for meeting times. It is always a good idea to take a digital camera if you have one, as well as a mobile phone.

## Risk assessment on a field trip

At GCSE you are unlikely to be involved in risk assessment, but you have some responsibility to stay safe during your visit. Here are some useful tips:

1 always work in groups
2 listen carefully to instructions, especially meeting times if you are away from the main group
3 take a mobile phone
4 always use common sense at all times.

More guidance on risk assessment can be found on pages 24–25.

### Tasks

1 Working with a partner, write a list of the likely safety risks associated with your planned field trip.
2 Explain how you plan to minimise these risks while on the field trip.
3 Explain how you will continue to monitor your own safety on the day of the trip.

## Some ideas for methodologies

The choice of methodology depends upon the type of study you are conducting. In this section you should provide an explanation of the following:

1 where you collected your data and why you chose these locations
2 how you collected your **primary data** and why you chose these methodologies
3 you should also describe any **secondary data** collection methodologies.

### Key terms

**Primary data:** new data that has never been collected before. If nobody has ever collected a particular piece of data at the same time and in the same place as you, it is primary data.

**Secondary data:** information collected from another source. For example, information in a textbook or on the internet would be considered secondary data.

### Liverpool: the field trip

*Case study*

For this study the students were undertaking a field study in an urban environment. The teacher gave them the following checklist for equipment:

■ waterproof jacket with hood
■ packed lunch
■ small amount of money
■ mobile telephone and digital camera
■ pen, pencil, ruler and notepad.

The teacher provided clipboards and data collection sheets that the students had designed at school.

The students agreed a range of methodologies before starting the investigation. Remember that they were interested to see if the environment the residents came from affected their perceptions of changes in the city. From their house price survey and census data search, they were able to identify four contrasting areas in the city of Liverpool. The teacher fully risk assessed the sites and knew that it was possible to visit all sites in one day. This made the trip viable. In each area, the students worked in groups and surveyed different parts of the study area. These were agreed before they left school.

To investigate their aim and research questions, they decided to focus on the following methodologies:

### Environmental quality survey

**Why?**

They chose this type of survey because they wanted to find out whether there was a link between the affluence of the area (measured by average incomes/employment type and average house prices) and the quality of the local environments.

**How?**

At each site, groups of three students were given areas for investigation by their teacher. The teacher had given each group a Global Positioning System (GPS) (see page 34) and reference numbers for the exact locations of their task. At each site, in groups of three, they agreed a score out of five for sense of space, traffic flow, landscaping and overall appearance. They carefully recorded their scores for each site they visited.

| Environmental quality | | |
|---|---|---|
| Feature | Scoring system | Study site number |
| Sense of space | 5 – Spacious<br>1 – Crowded | |
| Traffic flow/cars parked | 5 – Very few<br>1 - A lot | |
| Landscaping | 5 – Greenery clearly present<br>1 – Built environment | |
| Overall appearance | 5 – Attractive appearance<br>1 – Undesirable appearance | |

### Quality/Decay Index

**Why?**

As with the environmental survey, they were interested to see if affluence affected the quality of the housing.

**How?**

At each site, they were given five houses to grade according to the Quality/Decay Index (QDI) shown on page 15. They gave a score out of 10. They worked in groups of three and had to agree the score before they moved on to the next property.

| Quality/ Decay Index | Descriptor |
|---|---|
| 10 | Immaculate paintwork/ windows/ brickwork. Building materials/style show care and thought. Design is interesting – detail added. Improvements in evidence/excellently maintained. Overall, building is aesthetically pleasing. |
| 5 | Average paintwork/windows/brickwork. Building materials/style functional. Design is basic – 'no frills'. No evidence of improvement/some maintenance. Overall, building is satisfactory in appearance – it neither adds to nor detracts from the landscape. |
| 1 | Very poor paintwork/windows/brickwork. Building materials/style unattractive. Design is unattractive. Building in a state of disrepair – in need of immediate attention. Overall, building is unsatisfactory in appearance – it is an eyesore. |

**Key term**

Sampling: is an important concept at both GCSE and A-level. We take a data sample when it is not possible to collect information on the whole population. However, it is important to follow certain rules when sampling. You can find out more about this on page 27.

### Questionnaire

**Why?**

The students were especially interested in the views of the local people about the 2008 Capital of Culture.

**How?**

They designed a simple questionnaire to gauge local opinion on the changes taking place in Liverpool. Here are examples of the sorts of questions they asked:

1 Are you from this local area?

If the answer was 'no' to this question, they discontinued the survey as they were only interested in the views of people from the community they were in.

2 What changes has the 2008 Capital of Culture brought to Liverpool?

3 Did you visit any of the events or venues? If so, which?

4 How has the 2008 Capital of Culture affected you?

5 How would you rate the effect of the 2008 Capital of Culture on your community?

As part of their systematic **sampling** technique (see page 27) they asked these questions to every third person who passed them on the street where they were collecting data. For the open-ended questions, they asked each interviewee if they could record their response using a Dictaphone. Not everybody agreed to this. However, many did agree and this gave another source of primary data for analysis later.

### Digital image analysis

**Why?**

This is an important technique, which if used well, can add a really good feel to the study as well as meeting important criteria in the mark scheme.

Visiting **www.yell.com/ukmaps/ home.html** the students were able to obtain oblique digital aerial images of the study areas. Look at how Walton contrasts with Childwall. Which do you think is likely to have a better environment quality?

Write a list of the contrasts between Walton and Childwall

**Tasks**

**Collecting secondary data**

1 Describe one or more methods of secondary data collection.

2 Explain why you used this secondary data in your study.

### How?

In each area, the students took photographs which they felt best represented the community, using digital cameras (see page 119). These were annotated back at school to show the characteristics and key features of each area.

### Secondary data search 1 – census data

### Why?

This data was needed to identify key characteristics of the study area.

### How?

Students completed this aspect of the study before they conducted their fieldwork. This helped them to build up a picture of each area's characteristics. Using the website **www.neighbourhood.statistics.gov.uk/dissemination** they obtained local information on populations, employment and health characteristics. They also looked at crime and average incomes. They selected appropriate information before summarising this data for their study.

### Secondary data search 2 – house price survey

### Why?

Some students planned to investigate the connection between house prices and affluence by comparing house price data in each area to incomes and employment data from the census. Others used this data to see if there was any link to perceptions about the changes in the city. For example, they were keen to find out if places with low house prices were more likely to contain people with negative views about the 2008 Capital of Culture.

### How?

Students either gathered data from local estate agents or used various websites such as **www.rightmove.co.uk** to obtain house price data for the locations where they were planning to collect their primary data.

Childwall in Liverpool, a generally more affluent part of the city. Most housing is privately owned with gardens front and back as well as garages and driveways. Are residents in this part of the city more likely to have a positive view about the 2008 Capital of Culture?

Housing in Walton tends to be older with a higher density. Within Walton there are some more affluent parts, with evidence of more recent improvements such as the addition of a garage. How are residents in this part of Liverpool likely to view the changes brought by the Capital of Culture?

## Tasks

### Preparing your write-up

These tasks should take place after your field study in preparation for your write-up.

1 Describe how you collected your primary data. In doing this you should:
   i describe each methodology
   ii explain your sampling technique (if you used one).
2 Explain why you chose these methodologies and how they helped you to achieve your overall aim.
3 Evaluate your methodology. In doing this, you should:
   i describe the strengths and weaknesses
   ii explain how the limitations might affect the validity of the results
   iii suggest improvements to the methodologies.

> **Tip**
>
> There is no set rule on the number of methodologies to use. However, it is essential to collect some primary data. It is also preferable if you draw upon some secondary data, though this largely depends upon the type of controlled assessment.

# Transforming your data

In this section of your study you will be required to put your data in order, often using tables. You will also be required to show that you can use appropriate techniques to display your data. This includes using graphs, maps and statistics where appropriate. Before you can proceed, it is likely that you will have to share your data with either the rest of the group or the whole class.

You have a word limit, so the more techniques you put in, the more description and analysis you will have to do later. The key here is to use appropriate techniques and skills. You should also use a mix of techniques. Pages of bar graphs will do very little for the quality of your study; it is better to use a range of techniques from chapters 2, 3 and 4 of this book. Your teacher will be able to help with this.

## Tasks

**Understanding the techniques you will use**

1 With help from your teacher, make a list of the techniques you will be using for the results section of your study.

2 Find the techniques you want to use in this book and list the page numbers for each technique.

3 Spend 20 minutes or so reading about how you can use each technique to transform your data.

## Liverpool study: transforming data

*Case study*

The students started by annotating some digital images. They pointed out key features and characteristics, importing their images from digital format into a word processing programme (see photograph of Walton below). They then decided to analyse their primary and secondary data for the four areas.

For ease, they numbered the areas follows:

- Toxteth (Area 1) – an inner city location experiencing recent regeneration.
- Wavertree (Area 2) – an inner city location with some more affluent areas.
- Walton (Area 3) – an inner city area, with higher than average deprivation and poverty.
- Childwall (Area 4) – an affluent suburb.

Their teacher suggested they look at correlation techniques (see pages 61, 100) between different variables, such as:

- Quality/Decay Index scores and environmental quality.
- Average incomes (census data) and environmental quality.
- Average house prices and environmental quality.

One student decided to construct a scatter graph (see page 61) to investigate correlation between average house prices and environmental score. He first had to share data with other groups to generate enough data for his graph. The data he collected is shown on the following page.

## Take it further

Find out more about annotating photographs on page 119.

Originally, these buildings were stables for working horses. Notice high arch door.

There is now evidence of vandalism, a sign of the area's social problems. The buildings have been secured to stop any further damage.

The buildings are traditionally brick built with a slate roof, dating back to around 1880.

The change of land use to workshop/storage has not been successful as the properties remain vacant and derelict.

The industry and farmland in this area have long since gone. These buildings have become a relic of the past

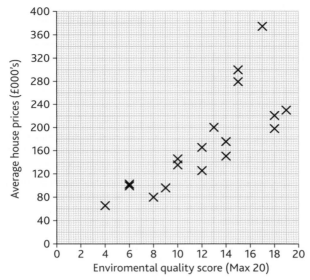

Scatter graph showing the relationship between house prices and environmental quality scores in the four study areas of Liverpool. What could be added to help analyse this graph?

Next, the students decided to plot some of the data on environmental quality (EQ) and Quality/Decay Index (QDI) on a sketch base map using the located proportional bar technique. More information on this technique can be found on page 46. Here is the data they used to create the map:

| | Environmental quality | Quality/Decay Index |
|---|---|---|
| Toxteth | 14 | 7 |
| Wavertree | 12 | 7 |
| Walton | 8 | 4 |
| Childwall | 18 | 9 |

Here is the data used to construct the scatter graph

| House prices (Areas 1–4) | Environmental quality score (max. 20) |
|---|---|
| 480,000 (4) | 20 |
| 198,000 (1) | 18 |
| 125,000 (2) | 12 |
| 375,000 (4) | 17 |
| 79,000 (3) | 8 |
| 101,000 (3) | 6 |
| 175,000 (1) | 14 |
| 145,000 (2) | 10 |
| 99,000 (3) | 6 |
| 95,000 (3) | 9 |
| 279,000 (4) | 15 |
| 299,000 (4) | 15 |
| 65,000 (3) | 4 |
| 165,000 (2) | 12 |
| 150,000 (2) | 14 |
| 230,000 (1) | 19 |
| 220,000 (1) | 18 |
| 199,000 (4) | 13 |
| 135,000 (2) | 10 |

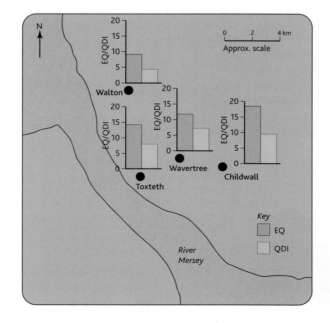

**Take it further**

If you can't find a suitable base map, another useful technique would be to draw a sketch map of your area. See page 36 for more on this.

Sketch map with located proportional bars showing EQ and QDI averages in the four study areas

Some of the students were encouraged to use a Geographical Information System for the display of census data. They used the Office for National Statistics website – **http://neighbourhood. statistics.gov.uk** – to obtain data on levels of unemployment and unskilled workers (Social Grade group E – see page 75 for more on this). By selecting appropriate information, they were able to display the pattern for Liverpool on a choropleth map (shown below) overlaying a base map of the city. They planned to use this data later in the discussion and analysis of findings.

Finally, the students decided to look at the questionnaire responses and analyse this information. The teacher gave the students plenty of choice about which aspects of the questionnaire responses to focus on. They first had to share data between different groups in the class. This was organised by the teacher. In terms of the overall aim of this study, the questionnaire data was crucial in helping to address the aim and research questions. Remember that this particular group of students wanted to see how different communities viewed the legacy of the 2008 Capital of Culture. The teacher gave the students a number of options. They could:

- quantify the views in each area on the number of new developments and what impact they have had on each area
- contrast the views on the impact of the Capital of Culture across the four areas
- investigate the number of visits to venues and events made by people in each area to see if there were any trends
- analyse the Dictaphone recordings for patterns in the views held by the people in the different districts.

### Key point

#### Using qualitative data from questionnaires

Qualitative data is very useful because it illuminates the perspectives of the local people. It gives much more information about attitudes and values. If you use extended responses in a questionnaire, you have to decide what material to include in your study. It is important to select views carefully. They should be representative of general opinion and not simply fit with what you want to show. You can draw on an unusual viewpoint or comment, as long as you make it clear that this view was only held by a minority. Like the rest of the controlled assessment, you have to decide what to include and what to leave out as you are restricted by your word count.

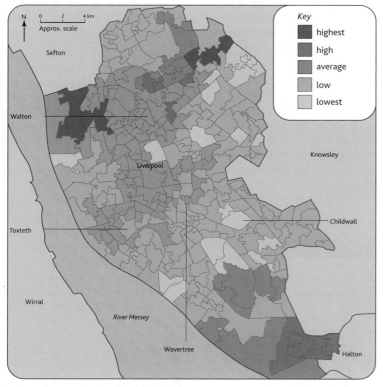

Unemployment rates in Liverpool – can you see any patterns?

# Analysis, discussion, conclusion and evaluation

You have to complete this part of your study under strict examination conditions. You will be able to use the rest of your study, but no other materials or textbooks. You will not have access to the internet unless this has been specifically authorised (for example by your examination board). It is very important to be well prepared for this section, and to know what sort of detail to include in each part of this high-control exercise, in order to maximise your marks.

## Discussion and analysis

Here is an extract of a student's analysis and discussion in relation to the above enquiry. This scored high marks.

*'Our findings showed that the environmental quality (EQ) in our four study areas was very varied as was the Quality/Decay Index of the buildings we surveyed in each area. For example, Childwall had an average EQ of 18 and a QDI of 9. In Walton the average EQ was just 8 and the QDI was 4. It was no surprise that house prices showed similar contrasts. I calculated the average house price in Walton to be £97,500 whereas in Childwall, it was £275,000 – a difference of £177,500. This is easily explained with an analysis of the types of jobs people tend to do in each community. The general pattern is that Childwall contains more higher-income professionals. This made sense because these people are the only groups who could afford the high house prices.*

*It was also interesting to analyse questionnaire responses in relation to these findings. I found people in Walton were generally less aware of the developments which had taken place, with the average person knowing of only three events during the 2008 Capital of Culture. This compared with an average of seven events in Childwall. This suggested people in Childwall visited many more events and venues than the people of Walton. It was also interesting to note how many people from Toxteth visited events and venues. Our questionnaire showed that on average people living in this community visited 12 events and venues during 2008. As Toxteth was so much closer to the city centre, I feel this is a possible reason why these groups were more likely to make more visits to different events in the city centre. Also, this area has undergone lots of recent improvement as gentrification has occurred. Like Childwall, this area contains lots of higher-income professionals ...'*

Can you see why this part of the study has scored high marks?

## Task

### Analysing the sample answer

1  Read the case study on this page and pick out examples of where this passage has:
   - identified patterns
   - offered interpretation
   - used data to support findings
   - analysed data by manipulating with additional calculations
   - identified and explained anomalies.

### Tip

In this section, you should bring together different aspects of your study. Describing your findings will only take you so far. You need to go further and pull together different parts of your findings. Higher-level skills include the following:

- identifying patterns and trends in your data
- offering potential explanations for the trends in your data (this is called 'interpretation')
- supporting these findings with evidence from your results
- manipulating your data by performing additional calculations from the data that you originally produced
- identifying anomalies within your data and offering potential explanations as to why these have occurred.

## Conclusion

In your conclusion you should simply revisit each of your research questions (or objectives hypotheses) and decide whether each one has been answered/achieved/proven. Deal with each one in turn and write a very brief summary to explain your position. Finally, look back at the aim and decide whether you have achieved it.

## Evaluation

Finally, you have to weigh up the strengths and weaknesses of your enquiry. You also have to consider how your results (and therefore conclusions) might be affected by the limitations of the study design and methodologies used. Make sure you do the following in the evaluation of your enquiry:

- assess the strengths and weaknesses of your methodologies
- explain what impact the limitations of the methods have had on the results
- examine the extent to which the conclusions might be unreliable
- suggest improvements to the study and further areas of investigation if you were to undertake the study again.

Here is part of a student's evaluation in relation to the above enquiry (this scored high marks):

*'Now I have completed the study, I have noticed a few issues which might have affected my results and therefore my conclusions. We did not really have a good system for sharing our data so I tended to swap data with people I was comfortable with. This meant that I did not get a representative picture of the whole of each study area. Also, while out on the data collection we had lots of disagreements about the EQ and QDI scores. Mostly we compromised, but sometimes we just gave different scores to each other and sometimes there were very big differences. So when it came to swapping data, results could be very different depending upon who you shared with. I would therefore say that this data was a little unreliable. However, once averages were taken to create our proportional located bar graph, these differences were evened out to an extent.'*

Can you see why this section scored high marks?

**Task**

2 Practise writing your analysis, discussion, conclusion and evaluation for your study. Make notes and try to remember what scores high marks in this section.

Using this section as your guide, you should be able to tackle your own controlled assessment. Don't forget that this is worth 25 per cent of your GCSE. Stay organised, listen to advice from your teacher, complete all sections to the best of your ability and you can be successful.

**Summary**

- The controlled assessment is worth 25% of your overall GCSE grade.

- It should be around 2,000 words though this does not include graphs, maps and tables of data.

- You have to concentrate under examination conditions. You should prepare for this in the same way as for any other examination.

# AS/A2 Fieldwork investigation

You no longer have the choice of submitting a written piece of coursework at A-level. Instead, you have to write about your fieldwork experience in an examination. Being well organised in preparation is essential. You have to undertake the fieldwork, write it up and then practise for the sorts of questions you might be asked in the examination.

This is quite a demanding task. Not only do you need to complete the work in the field, but you to have to think very carefully and clearly about what you are doing and why you are doing it.

## Why do fieldwork?

Fieldwork is a requirement of all geography courses, and you will be examined on it at both AS and A2. Your teacher will set up the fieldwork experience, but it is important that you are involved in all aspects of the planning from the start.

Fieldwork is an important part of understanding how the geography you learn in class applies to the real world. One of the things that you usually learn is that the geography of the real world is very much more complicated than the geography of your textbook. It is also great to get out of school and think like a geographer for a whole day.

## What about the examination?

The following table will help to prepare you for the sorts of questions you might face in an examination. Most geographers agree that a fieldwork enquiry has clearly defined sections or areas, irrespective of the type of study. From this it is possible to summarise the areas you might be asked questions on in an examination. However, you will also need to practise with past paper questions from your own specification if you want to be ready for any examination that is testing your knowledge of your geographical enquiry.

Fieldwork investigation is a requirement of all geography courses

---

**In this section you will learn:**

1 the key elements of an A-level fieldwork enquiry

2 how risk assessment is carried out

3 how the fieldwork is examined

4 the tips and hints for scoring high marks in the examination.

---

**Tip**

You can be questioned on any aspect of your fieldwork, so you need to experience all aspects of it.

---

**Tip**

**Key questions in A-level fieldwork**

- Do you know why you are doing the fieldwork?
- What are the underlying theories, ideas or issues?
- What is the sequence of the investigation?
- Can you describe what you did?
- Can you explain what you did?
- Can you justify all of your choices?
- Can you summarise your findings?
- Can you evaluate your study?

---

**Tip**

**Deciding on the aim of your study**

The aim of the study should be based on some area of the relevant part of your examination board's specification. Your teacher is already likely to have decided, though you may create an aim through group dialogue.

## The nine stages of a fieldwork investigation

| Area of your enquiry | Possible focus in the examination |
|---|---|
| 1 Aim (including objectives, hypotheses or research questions) | Describe the aim.<br>Link the aim to your theory. |
| 2 Location for study | Suggest why this area was suitable.<br>What risk assessment was carried out in the study area to minimise potential hazards during primary data collection? |
| 3 Underpinning theories or processes (and links to your specification) | Describe underlying theory or process and show how this links to your study. |
| 4 Methodology (including sampling) | Describe the methodology.<br>Describe the sampling techniques.<br>Justify the methodology.<br>Discuss strengths and weaknesses of the methodology. |
| 5 Results | Describe your findings.<br>Describe a technique used to present results.<br>Evaluate the usefulness of the technique for displaying data.<br>Suggest an alternative technique for displaying data. |
| 6 Discussion, analysis, conclusion | Discuss your findings.<br>Describe your methods of analysis.<br>Explain how far your findings matched the original theory.<br>Outline the conclusions. |
| 7 Evaluation | Measure the strengths and limitations of the study.<br>Suggest improvements. |
| 8 Further research | In the light of findings, suggest further research. |
| 9 Use of new technologies | Explain how you used modern technology in your enquiry and evaluate its effectiveness. |

# Background preparation and research

It is important to obtain some background information on your study area. Look for factors that explain the context of your local area and provide information that might be useful later when you are explaining your findings. Also, obtain a suitably scaled map to locate your study area.

The research you do largely depends on the type of study you are undertaking. At A-level, you will need to be able to demonstrate that you understand the theory, idea or concept that has formed the basis of your study. This material can be found in many forms, but should in some way be linked to your specification. Here are some possibilities:

■ a theory from one of your A-level textbooks

■ a model that you might be testing for its validity

■ a recent news report about an issue in your local area.

Whatever the focus is here, the key for you is in being able to link the background succinctly to the aim of the study.

> **Tip**
>
> Some model testing approaches to fieldwork are a little dated. Some geographers argue that testing the validity of urban models such as the Burgess concentric ring model is now outdated. This is because the Burgess model does not 'fit' the typical British city and was never designed to do so.
>
> Models such as the Bradshaw model in river studies can still provide the basis of a valid, up-to-date approach to fieldwork. This is because many British rivers 'fit' the model's key assertions, i.e. the model is still valid.

## Investigating the changing characteristics of the River Alyn, North Wales

As part of their AS geography studies, a group of students decided to investigate the extent to which the Bradshaw model could be applied to the River Alyn, a river in their local area.

The students decided to use hypotheses for the basis of their enquiry:

1 Velocity increases with distance downstream.
2 Discharge increases with distance downstream.
3 Channel efficiency increases downstream.
4 Load becomes more rounded downstream.

Can you describe the key suggestions of the Bradshaw model?

### Tasks

With reference to your own A-level geographical enquiry:

1 i State the aim of your enquiry.
  ii State two objectives, hypotheses or research questions designed to help you achieve your aim.
  iii Explain how these will help achieve your aim.

2 Describe your study area and explain why this site was chosen for this enquiry.

3 Outline the background theory that has shaped this enquiry.

## Using objectives, hypotheses or research questions

You have three choices here. Whatever type of study you are doing, you will need to break down the enquiry into manageable components that permeate the study.

### Objectives

These are used to set out specific 'mini' aims that will help to achieve your overall aim. Like the other approaches, you may use between one and four objectives, depending on your study. You may also choose to use objectives as well as hypotheses or research questions. This really depends on your study and your own preferences.

### Hypotheses

There are statements that you test and then reject or accept. You would usually have between one and four hypotheses, depending on the type of study.

### Research questions

By answering the research questions you pose, you achieve the aim of the study. Again, there are usually between one and four research questions in a typical A-level enquiry.

### Tip

**Carrying out a risk assessment**
Before you go anywhere, your teacher will have risk assessed the site you are planning to visit. However, you also need to be involved in this process at A-level. Risk assessment involves looking at the site, its surroundings, the weather and possible hazards. You have to show that you understand the obvious possible dangers, and that you know how to reduce the likelihood of the risk becoming a real hazard. Finally, you need to show that you are continuously monitoring the risks as the day progresses.

## Tasks

**Risk assessing in practice**

Look at the image (below) of a group of students undertaking a beach study in their local area. On a suitably sized photocopy, complete the following activities:

1 Label the photograph to identify possible risks. Using a sharp pencil, make sure that you clearly point at the place where you can see the risk.

2 Using a different-coloured pen or pencil, annotate the photograph to explain how you might minimise each risk.

3 Write a short paragraph that discusses the ongoing risks, during the day, with this sort of fieldwork.

4 Apply this approach of fieldwork risk assessment to your piece of fieldwork.

Stages in risk assessment

### Tip

If you need further help with photograph annotation, turn to page 118.

What risks are involved in this fieldwork study?

# Equipment checklist

Your teacher will provide all the technical equipment. For a river study, you are also likely to need your data recording sheets, wellington boots or waders, clipboard, waterproof clothing and packed lunch. Ideally you will be provided with a checklist a few days before you go out.

Here is the list of equipment used in this study:

- Ranging poles
- Stopwatch
- Tape measure
- Cork (or flow meter)
- Powers' Scale of Roundness cards
- 30 cm ruler and a metre ruler
- Long piece of rope or chain
- Data recording sheets.

# Methodologies

The key to success here lies in thinking ahead to your examination while out in the field. For the methodology, you may have to be able to explain how you collected primary and/or secondary data (see page 13). You may also have to justify your methods (see command words in Appendix 2) in relation to your original aim. You may also be required to link primary data to risk assessment and sampling (see page 27 for sampling methods).

## Case study

The students had to agree 10 sites along the course of the river for their data collection. These were agreed before they left and all students were involved in the risk assessment (see pages 24–25). They decided to focus on five methodologies in order to test their hypotheses. Here is a brief summary of what they did:

1 Channel width was measured by pulling a tape measure tightly across the channel from bank to bank at the level of the water. The bankfull width was also measured using the same technique, but this time from the top of the bank on either side.

2 The channel depth was measured at regular intervals across the channel. For narrow sections, near the source, intervals of 20 cm were used. For wider sections downstream, intervals of 50 cm were chosen. A metre ruler was placed vertically down into the water and the depth was measured in centimetres. The bankfull depth was also calculated in the same way, but a piece of string was held across the channel from the top of each bank and depth was calculated from this point.

3 The wetted perimeter was measured by placing a long piece of rope in the river from the water level on one side of the channel to the other, making sure the rope was in touch with the bed and banks at all points across the channel. The length of the rope was then measured as accurately as possible.

4 To calculate the velocity, a cork was placed into the middle of the river at the start of a 10-metre section in the river. A stopwatch was then used to time how long it took for the cork to travel 10 metres. Three samples were taken and the average of the three readings was calculated later.

5 Finally, 10 pieces of bedload (clasts) were sampled in the middle of the river at each of the 10 sites. This was done by collecting the nearest 10 pieces of bedload to the mid-point of the channel. Powers' Scale of Roundness (below) was recorded for each clast.

Photographs were taken at each section and field sketches drawn of the shape of the channel.

### Key term

**Bias:** occurs when a particular piece of information is intentionally selected or is more likely to be selected because of the sampling technique. For example, it may be desirable to interview only females in a study investigating perceptions of crime on a local community. This study would have bias built into the questionnaire, males would be disregarded. Unless you are intentionally seeking to identify a biased set of data, you should design a sampling method that avoids bias.

### Tip

For secondary data, recent rainfall levels were investigated using the Met Office website (www.metoffice.gov.uk). They were interested to find out their impact on discharge levels.

| Very angular | Angular | Sub-angular | Sub-rounded | Rounded | Well rounded |
|---|---|---|---|---|---|

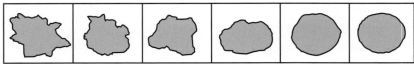

Powers' Scale of Roundness

# What is the best approach to sampling?

You should conduct your sampling in a way that will best represent the data you are collecting. There are three main types of sample: random, systematic and stratified.

### Random sampling

In this technique, every item has an equal chance of being selected. In this sense, there is no **bias**. For most studies this is the most desirable approach, though you may specifically want to identify a particular subset within the data and sample this. In this case you should use a stratified approach (see below).

Random sampling gives every piece of data in a parent population an equal chance of being selected

The most likely way of using the random sample is to use a random number table. You might use this technique if you wanted to select a particular grid reference, for example. Another simple way of random sampling is to draw numbers out of a bag. For example, if you wanted to make a questionnaire truly random, you could place numbers from 1 to 50 in a bag; you could then draw numbers out to identify the individuals you would like to stop and question. Remember to replace the numbers in the bag, so that everybody has an equal chance of being selected. Most scientific calculators can also generate random numbers.

### Systematic sampling

This approach is different from random sampling, in that there is some structure or underlying order to the way in which data is selected for sampling. Using the questionnaire example, interviewing every fifth person would be a systematic approach.

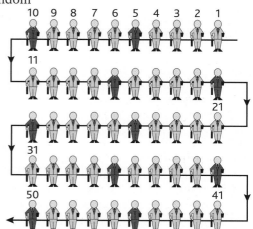

What sampling approach has been used here?

### Stratified sampling

You would choose this approach if you wanted to bias your data. For example, if you knew the proportion of pensioners in an area was 30 per cent and you wanted to represent this in your sample, you would make sure that your sample contained pensioners' views to the value of 30 per cent of your sample.

### Line, point and area sampling

Deciding whether to use line, point or area sampling depends upon the sort of data you are collecting. For example, you would use line sampling if you were gathering data along a transect (see page 85). You would use area sampling if you were conducting a quadrat sampling technique (see page 85). You would use a point sampling technique if you were interested in sampling data over a wide area.

This sampling technique above uses a combination of systematic and stratified approaches. Imagine this area represents different land uses on a farm. Larger areas are sampled in greater frequency and vice versa

---

**Tasks**

**Describing your methodology**

1 Using your enquiry, describe how you collected primary data and secondary data.

2 Explain how your sampling technique ensured data collection was reliable.

3 Justify your methods in relation to the aim of the enquiry and background research.

4 Evaluate the methodology for primary data collection.

# Presenting your findings

In this section you are required to organise your findings and use suitable presentational techniques to show your results. Using tables of data is a good way of organising your findings, but you also need to transform this data using a variety of techniques from chapters 2, 3, 4 and 5. Remember that in the examination you may be required to:

- describe a technique you used
- explain why you used this technique
- discuss the strengths and weaknesses of the technique
- suggest an alternative technique and explain why you rejected it
- outline how new technologies can assist with the presentation of findings.

**Case study**

The students decided to display their findings using a variety of techniques. Here is summary of how they decided to display their data:

1  Scatter graphs (page 61) were used for investigating correlations between variables such as pebble size, discharge, velocity and distance downstream.

2  Located proportional circles were used to display the estimated pebble volumes.

3  Line graphs (page 77) were used to draw cross-sections of the river at various points along its course.

4  Because they had 10 pairs of data, a Pearson Product Moment Correlation Coefficient test could be carried out on the data (see page 103). They decided to examine the statistical relationship between the velocity and discharge data. Before they did this they had to decide whether the data was normally distributed (see page 92).

**Tip**

If you are faced with a question in your examination about methodology or results, make sure you choose something about which you can write a great deal. For example, there is far less to write about bar graphs than located proportional circles.

**Tasks**

**Describing a technique**

1  With reference to a technique you used to display your findings:
   i  Draw a sketch showing how you used the technique.
   ii  Describe how you used the technique.
   iii  Discuss the strengths and weaknesses of this technique for displaying your data.
   iv  Suggest an alternative technique and explain why you rejected it.

2  Outline an approach whereby new technologies could be utilised for the display of data.

It is important that you present your data using the most suitable technique

# Discussion, analysis and conclusions

Just as with controlled assessment (page 9), you are expected to describe and analyse your results. You should also try to make use of the findings, by offering explanations for the outcomes. It is also important to identify any anomalies in your study, and even better if you can explain these unusual occurrences within your data. At A-level, you would be expected to have used some statistical analysis and this should also figure in this section of your study. The statistical test above is therefore both a tool to display findings but also analyse them.

Your conclusion should briefly revisit the original aim, objectives/hypotheses/research questions and outline your position in relation to these elements.

# Evaluation

This may appear throughout your study or be in its own dedicated section (usually at the end). As with the controlled assessment, you should discuss the strengths and limitations of different aspects of the study, but particularly how the methodologies affected the validity of the results and reliability of the conclusions. Remember, of course, that it is in the examination that your knowledge will be tested and so you should try to recall key elements questioning methodology, results and conclusions. It is also useful to suggest improvements to the study. This may involve different lines of enquiry or extensions to the existing arrangements.

> **Tip**
>
> In the examination, when writing about the findings of the study or evaluating the effectiveness of the enquiry in meeting its aim, it may be useful to draw on actual evidence and data gathered throughout the enquiry process. Do plenty of revision to make sure you have data to draw upon. You are only likely to have between 30–60 minutes (depending on the paper) to express all your ideas, so be succinct and draw on evidence gained in the field.

## Tasks

**Evaluating your study**

1 Summarise the findings of your study.
2 Describe to what extent the conclusions meet the aim of your study.
3 Measure the strengths and limitations of the study.
4 Suggest further areas of investigation.
5 Describe how modern technology was used in either your fieldwork or the subsequent write-up.
6 Evaluate the effectiveness of the use of new technology in your study.

## Sample answer

*Case study*

In this evaluation the candidate has shown that they understand the limitations of the study and how these have affected the outcomes. There is also an attempt to suggest improvements:

'Our river study was flawed because we only visited on one day. There was not enough time to re-test some of the methods, such as velocity measurements. For example the cork kept on getting trapped in small crevices in the upland section near to Site 1. We are sure that this gave unreliable readings. When this data was transformed back at school, misleading relationships were suggested. We really needed to perform the data collection a greater number of times. We could have also compared our study with another river in a different drainage basin to see if the relationships turned out the same.'

#  The Extended Project Qualification

For you, as a geographer, this qualification represents an opportunity to obtain credit for studying an area of interest at a deeper level, counting towards your Uniform Marks Scheme (UMS) points for university. When applying to university you will need to show that you have something different or extra to offer a university department. The EPQ could be the solution. Even if you are not planning to go to university, the EPQ will still add something distinctive to your CV which employers are sure to find interesting. Before continuing, you first need to check that your school offers the EPQ.

## How to complete an EPQ

All the major examination boards offer the EPQ, so you need to be familiar with specific guidance offered by your own board. The common elements will be considered in detail over the next few pages. Here are some key questions and answers about the EPQ:

| Question | Answer |
|---|---|
| What can I do my EPQ on? | You can choose from four areas (see overleaf). For a geography project, you would realistically be restricted to either a dissertation or an enquiry. |
| How much writing is involved in the EPQ? | This should be around 5,000 words for a piece of geography coursework. |
| Do I have to do the EPQ on my own? | Yes and no. The EPQ is an independent project, so you have to write the project up on your own. However, you will also receive guided learning hours, i.e. lesson time of up to 120 hours. |
| How is lesson time used? | Your tutor will guide you through the project from start to finish. This will involve specific help in structuring the project and managing your time. Also you will get help with personal, learning and thinking skills (PLTS). These are the 'tools' you need to complete the project (see page 32). |
| Who will mark my EPQ? | Your tutor will also be your assessor. In the first instance, your tutor will mark your project. If there is more than one tutor, your work may be marked twice. This is called 'standardising'. A sample will then be sent off to the examination board to check the marking against national standards. |
| Why should I study for an EPQ? | You will gain an enriching experience by studying something in which you have a genuine interest. Also you will get important UMS points, as this qualification is worth half an A-level. If you are going to university, this could give you the edge and it will certainly impress a future employer. |

The first thing you have to do is choose an area of interest. This need not be geographical. It can be any area of school life or even an interest you have outside school. When considering an EPQ using a geographical theme, you have a wide range of types of project to choose from:

| Type of project | Involving | Geographical example |
|---|---|---|
| Dissertation | This would form an extended piece of writing to a maximum of 5,000 words. | Researching evidence of enhanced global warming. |
| Investigation/enquiry | A written structured investigation including the collection of primary/secondary data. | Any A-level enquiry could be written up in full and submitted as an EPQ. |
| Performance | Demonstration of practical skills and evidenced in a performance. | Not really suitable for a geographical topic or theme. More likely to be based on a theme from performing arts or physical education. |
| Artefact | A practical construction exercise relating to engineering, art or ICT. | Could link to geography through a cross-curricular theme such as sustainability. Could be a painting, sculpture, web design or engineering construction. |

**Key point**

If you are planning to begin an EPQ on a geography related theme, read pages 22–29 on how to set up and write up an A-level enquiry, and pages 127–134 on conducting research in geography.

## Tasks

**First steps to take towards an EPQ**

1 Find out which examination board your school uses for the delivery of the EPQ:
   i go to the appropriate website. For example, for AQA you can visit **www.aqa.org.uk**
   ii find the specification for the EPQ
   iii read the pages relating to the course requirements.
2 Draft a project title and aim.
3 Produce a brief summary of your EPQ. Include the following in your summary:
   i an outline of the plan, including dates for completion of specific aspects
   ii research opportunities
   iii data collection methods.
   You have now completed Part 1 of your project proposal.

**Key points**

**Level 2:** refers to study at Key Stage 4. For example, you can study for a BTEC at Level 2 and GCSEs are also Level 2.

**Level 3:** refers to study at Key Stage 5. For example, the Diploma is available as a Level 3 qualification as are A-levels. The Extended Project Qualification is also available as a Level 3 qualification.

# Personal, learning and thinking skills (PLTS)

Whatever theme you choose for your EPQ, you will need to show evidence that you have developed the following personal, learning and thinking skills:

# The functional skills

You will also be expected to demonstrate a range of functional skills throughout the development of your EPQ.

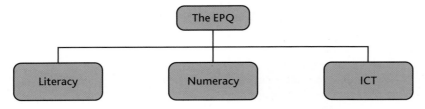

## Tasks

4  Visit the following websites:
   i  www.qca.org.uk/libraryAssets/media/PLTS_framework.pdf (PLTS)
   ii www.qca.org.uk/libraryAssets/media/QCA-07-3472-functional-skills_standards.pdf (Functional Skills)

5  Write a brief report of 500 words that explains how your project addresses both PLTS and the functional skills.

You have now completed Part 2 of your project proposal and should submit this to your tutor, who will comment on the suitability of your plan.

# The next stages...

Once you have had your theme approved by your tutor, you can get started. There are important elements of the EPQ beyond the writing up of the project. Here are some tips and advice:

■  Stick to the timeline you have submitted. Deadlines creep up really quickly. You must stay organised and on top of your studies. It is likely that you will have other study commitments and you have to balance these with your work on the EPQ. Unlike your other courses, it is really up to you to organise yourself.

■  You must keep an ongoing record of activities. You will be expected to keep a log of the various stages of the project. This is part of the assessment, so keep on top of it. Your tutor will give you more guidance on this.

■  You are expected to use new technologies as part of the EPQ. For geography, this could come in the form of Geographical Information Systems (see page 123).

- You may come across other forms of evidence as you work through the project. This should also be kept and submitted. It might include photographic evidence, primary data collection or presentation slides, for example.

At the end of the project you must present your findings to a non-specialist audience. Your tutor will decide the format of this. For example, you may have to present your findings to the other students in your EPQ group or you might just do a one-to-one presentation with your tutor. Evidence has to be kept of this presentation.

At the end of your project you must present your findings to a non-specialist audience

## How the EPQ is assessed

At the end of the project, you will obtain a mark in the range of grade A*–E. Failure to meet the requirements of a grade E will result in a U (unclassified). However, as there is no written formal examination, you will know well in advance of the final deadline if you are likely to fail.

Once complete, your work will be marked against four Assessment Objectives (AOs). These are the same for all examination boards, though mark allocations for each section do vary a little.

| Assessment Objective | Percentage of course | Features |
|---|---|---|
| Manage | 17–20 | Identify, design, plan and complete the individual project, or task within a group project, applying organisational skills and strategies to meet stated objectives. |
| Use resources | 20–22 | Obtain and select information from a range of sources, analyse data, apply relevantly and demonstrate understanding of any appropriate linkages, connections and complexities of the topic. |
| Develop and realise | 40–44 | Select and use a range of skills, including new technologies, to solve problems, to take decisions critically, creatively and flexibly, and to achieve planned outcomes. |
| Review | 17–20 | Evaluate outcomes including own learning and performance. Select and use a range of communication skills and media to convey and present evidenced outcomes and conclusions. |

You now have everything you need to get started on your EPQ. Stay organised, listen to advice from your tutor, meet your deadlines and you can be successful.

## Using an atlas

As with many other skills, students are generally using atlases much less frequently than in the past. To an extent this is understandable. You can now get a map of anywhere in the world in seconds using the internet. However, there are still some important skills to develop through the use of atlases, such as using the **latitude** and **longitude** system of location referencing.

Have you ever wondered how a Global Positioning System (GPS) works? These are found in most modern cars and are often referred to as satellite navigation systems (or 'Sat Nav' for short). A GPS uses several satellites (orbiting around the world) to pinpoint your location on the ground. It does this by working out your exact latitude and longitude on the earth's surface.

Lines of longitude are like lines dividing segments of an orange. They do not run parallel to each other like lines of latitude. Instead, they start and finish at the Poles. They are measured in degrees, with the **prime meridian** being 0°

Lines of latitude are imaginary lines that are numbered using a system of degrees. The **Equator** is located at 0°. The North Pole is 90°N and the South Pole is 90°S

When the world is drawn flat on a map, it is easier to see the lines of latitude and longitude. They operate together in a similar way to grid lines on an Ordnance Survey (OS) map. Read the section on grid referencing (page 39) before continuing. The main difference is that an OS map uses eastings followed by northings. With an atlas, the latitude measurement is given first, followed by the longitude.

You should be able to see that on the map on the next page, 30° intervals are used between each line. However, as the world is a very big place, these intervals are not particularly helpful in locating specific places such as cities or even small countries. So, instead of saying that India is 20°N and 80°E, we might need to be much more specific in identifying a particular city in India or perhaps just part of that city. Satellite navigation needs a system for identifying a pinpoint location. Clearly, just using degrees is not enough.

### In this section you will learn:

1 the link between atlas skills and Global Positioning Systems

2 the difference between latitude and longitude

3 how to find places using an atlas.

Satellite navigation uses a combination of GPS and road maps to get you to your destination

### Key terms

The **Equator** is the most well-known line of latitude. It is a line drawn horizontally around the middle of the planet. Other well-known lines of latitude are the Tropic of Cancer and the Tropic of Capricorn. Use a web search to find out what they represent.

The Greenwich Meridian, (or **prime meridian**), and the International Date Line are the best-known lines of **longitude**.

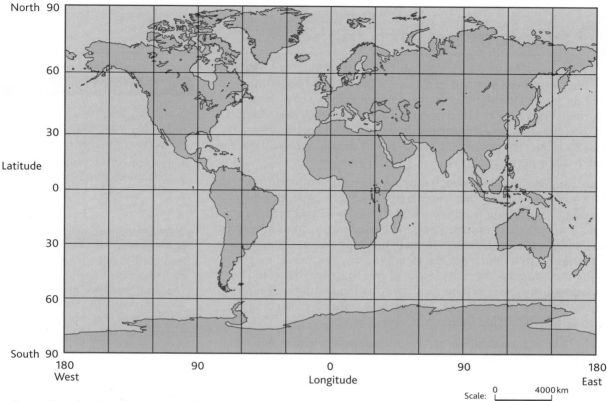

Atlases therefore have a system of using degrees, minutes and seconds. Each degree has 60 minutes and each minute has 60 seconds. The location of Cairo in Egypt would be:

30°3′24″N  31°15′36″E

This is now enough detail to locate Cairo accurately on a map showing this system.

## Tasks

1  Using the map above, estimate the latitude and longitude for the UK using degrees alone.
2  Using an atlas, check your answer.
3  What conclusions do you draw from this?
4  Practise using an atlas by locating the following places:
   Rio de Janeiro, Berlin, Melbourne, Johannesburg.
5  Name the nearest major settlement to each of these cities and give its exact location using degrees and minutes.

### Extension research task

6  Explain how lines of latitude link to temperature and seasonal change.
7  Explain how lines of longitude link to time.
8  Using an up-to-date atlas, create a country profile for a case study of your choice. Include information such as population, climate, farming, industry, etc.
9  Using the same country as in question 8 above, create a country profile from an internet site such as The World Factbook – www.cia.gov
10  Compare your country profile from your atlas with your country profile from The World Factbook (which can be downloaded from the above website). What conclusions do you draw?

## Tip

Many atlases contain a wealth of reference information on different countries around the world. If you are starting a new research topic on a city or country you have never studied before, see what you can find out from an atlas, even before you do a web search.

## Key point

Most atlases only use degrees and minutes.

# GCSE AS/A2 Creating a sketch map

If you have ever been stopped in the street by somebody needing directions, you may well have drawn a sketch map to show them how to get there. This type of sketch map is often referred to as a '**mental map**', because it has been constructed entirely from your memory. It is unlikely to be very accurate, but if it is well drawn, it should get the person to their destination.

You are also likely to use a sketch map for presenting other forms of data. In other words, your sketch map becomes a **base map** for displaying primary data from your field study. These sketch maps are useful for geographers in that they can be used to get rid of unwanted items and information from an OS map, allowing you to display only the things that matter for the task. Whereas the mental map showing road directions is not usually drawn to scale, you would be expected to draw a sketch map for displaying field data (base map) to a reasonably accurate scale. You can see how a sketch map could be used to display field data from a study on page 18.

If you have an accurate map (at the right scale) from which to draw your sketch map, the task is much easier as you can simply trace the aspects you need, provided you have the right paper. Often though, you have to create your sketch map by simply looking at a map, choosing the things you need to focus on and then sketching these on to a piece of paper.

Some tips:

- Use a box to keep the map tidy and professional looking.
- If it is a town centre sketch, start in the middle and work outwards – you are less likely to get the scale badly out of proportion.
- Add only the essential information for your project.
- Add an approximate scale (get help with this if you are not sure).
- Always add a title and a north arrow to show its **orientation**.

Let's look at an example from human geography.

## In this section you will learn:

1 why sketch maps are important
2 when to use a sketch map
3 how to draw a sketch map
4 examples of where you might use a sketch map in your studies.

## Key terms

**Mental map:** a sketch map created directly from memory. This is a difficult skill and it is almost impossible to draw to an accurate scale. If drawn by a local person, they tend to focus on local landmarks.

**Base map:** a map that is used for the display of other information, such as data collected on a field trip. These maps usually have very little on them to start with, showing only the important information for that study.

**Orientation:** the way the map is facing. Usually the map will be orientated with north at the top of the map. However, this may not suit you. Just make sure you show which direction north is on your sketch map.

## Case study

### Measuring pedestrian flows in a central area

A group of students were interested in gathering pedestrian density figures in order to help them identify the central area of their local town centre, Altrincham, a small market town south of Manchester. They wanted to use part of Altrincham as a base map to create an isoline map showing changing pedestrian densities across the central area. First though, they decided to draw a sketch map of the town centre because they could not find a suitable base map (showing only the roads) to display the data. They began looking at a 1:25 000 OS map. They used this to identify their study area. They also looked at Google Earth to get an accurate satellite view of the image they wanted. They used both these sources as well as their own knowledge to create the sketch map.

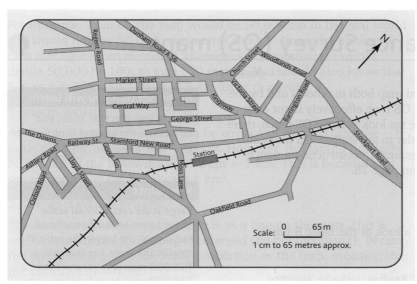

Can you see the advantages of a sketch map compared with the other options below?

A satellite image of Altrincham from Google Earth – a Geographical Information System. Here the image uses a combination of road mapping and satellite technology

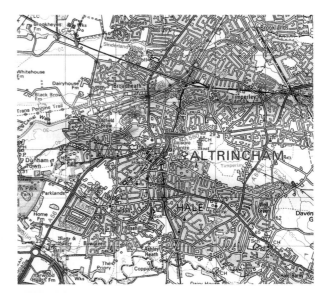

A 1:25 000 map extract of Altrincham.
Notice how the orientation of this map is different from the satellite image on the left. Can you list the differences between the two figures?

## Tasks

1 Obtain a 1: 25 000 map of your nearest town centre. If you live in a large city, you will need to focus on one small area. If you cannot get a map, use the internet site **www.ordnancesurvey.co.uk/oswebsite/getamap**, where you can get a free electronic map at 1:25 000.

2 Draw a sketch map of your central area.

3 Add a box around your map and give it a title, key and some indication of the scale.

4 Describe the difficulties of creating a sketch map in this way.

## Summary

- Drawing accurate sketch maps takes practice.
- Keeping everything to the same scale is difficult.
- Choosing the best orientation for your task is important.
- Deciding what to add and leave out is vital.

# The key

The key is another crucial aspect of reading OS maps. Always remember that the key is simply trying to take the real world and make its features recognisable to you on a map. The key is a combination of common-sense symbols and letters on OS maps. For example, a caravan site is a tiny picture of a caravan, a public house is shown as PH and so on. Examples of some other symbols are shown on the right (also see Appendices 3 and 4).

## Height on maps

There are three main ways of showing the changing shape of the land (relief). The three ways are layer colouring, spot heights and contours. Layer colouring is not used on OS maps. It is a very effective system of using lighter and darker colours to show height changes and is often used in atlases.

Spot heights show heights in numbers. These are usually found at the tops of hills and mountains marked with a black dot or small triangle (triangulation point).

On OS maps, height is most commonly shown using contour lines. These are faint brown lines that join places of the same height. Height is always given as metres above sea level. Contour lines are drawn at intervals of 10 metres on a 1:50 000 map. At 50-m intervals, the contour line is a little darker brown. To help you get started, you need to understand that closely packed lines mean the land is steep and widely spaced contours means the land is gently sloping. Have a look at the OS map extract below and see what you can identify. You should see that the shape of the land has a big influence on how the land is used.

Some of the symbols found on a 1:50 000 OS map

> **Tip**
>
> **Is it human or physical?**
>
> When you look at an OS map, you should quickly see that it is dominated either by physical features or human features. In an examination, an OS map will have been carefully chosen by the examiner. Identify key features as quickly as possible.

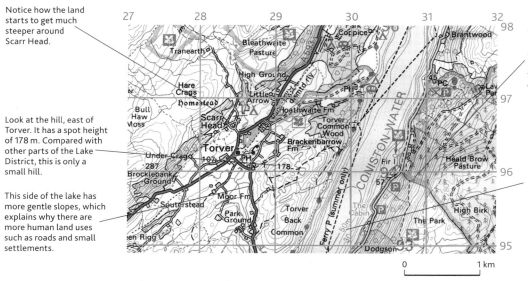

Notice how the land starts to get much steeper around Scarr Head.

Look at the hill, east of Torver. It has a spot height of 178 m. Compared with other parts of the Lake District, this is only a small hill.

This side of the lake has more gentle slopes, which explains why there are more human land uses such as roads and small settlements.

This side of the lake is much steeper. This explains why there is little human activity and the land has been given over to forestry.

Contour lines can also show depth of water. Coniston Water is 50 m deep. Notice how the contour lines are closely packed. This lake becomes deep quite quickly.

Coniston Water, Cumbria

## Tasks

Using the 1:50 000 map extract below of Stratford, East London:

1 How many churches are located in grid square 3482?

2 Name two of the grid squares in which Leyton is located on the map.

3 What is the direction from London Fields train station located at 348843 to the hospital in Hackney, grid reference 355853?

4 To the nearest half kilometre, how far is it from London Fields train station to the bus station in Stratford?

5 Calculate the length of the path around the perimeter of Victoria Park.

6 Suggest a possible reason why there is little development around the recreation grounds in the north of the map.

7 Plan a journey from the train station west of Leytonstone (383875) to the hospital north of Hackney (355853). Describe the journey. Include distance, direction and landmarks.

### Extension

8 The main site for the 2012 Olympic Games is Stratford (grid square 3884). Using evidence from the map below and your own knowledge, explain why this is a suitable site for the Games.

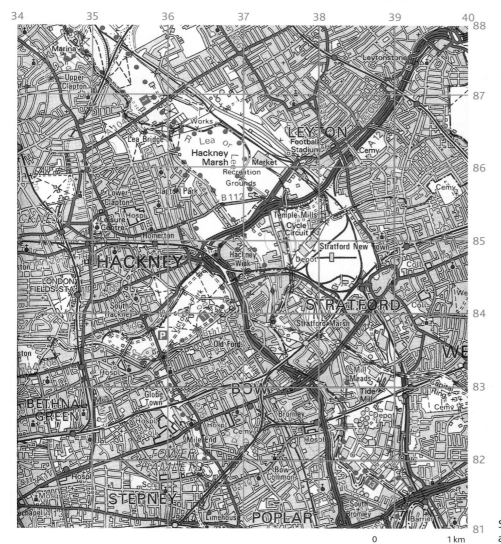

0    1 km

Stratford, East London – an area of dense population

# Showing movement on maps: flow, desire and trip line maps

## Flow line maps

A flow line is used on a map to show the actual flow and direction of something (for example, traffic). The flow line is drawn proportional to the number travelling along the route by the use of a suitable scale.

Flow lines are more likely to be used in large-scale maps, e.g. where roads, motorways, railways are actually wide and long enough to display the information. The information is usually displayed as actual values and is often given as a rate (e.g. 50 cars in 10 minutes). Percentages can also be used, perhaps to compare flows in opposite directions. For example, you would expect more vehicles to enter a town centre in the morning than would be leaving.

> **In this section you will learn:**
>
> 1 the difference between flow, desire and trip lines
> 2 when it is appropriate to apply each technique
> 3 how to create flow, desire and trip line maps using primary and secondary data
> 4 how to interpret the map
> 5 key points about the use of each technique.

### Case study

#### Measuring traffic flows in Manchester

A group of GCSE students decided to investigate the likely impact of the planned **congestion charging** scheme in Manchester as part of their controlled assessment (see page 9). They were interested to see how the flow of traffic might change as a result of Manchester bringing in a congestion charging scheme similar to London's. After gathering their data, they planned to investigate how traffic might decrease. They knew that the number of vehicles on London's roads dropped by 30 per cent in the first six months after the Congestion Charge was brought in. Although the scheme planned for Manchester was different, they used this figure to estimate changes to the number of road users.

*Methodology*

After a full risk assessment, students worked in pairs counting vehicles using a simple tally chart (see page 88). First, they agreed locations (choosing road junctions, as it is easier to count cars at a junction) along all major routes into and out of the city. Next, they synchronised their watches before leaving and counting vehicles for 30 minutes, starting at 9am on a weekday. They chose to display their findings using a flow line map (shown on page 43).

> **Key terms**
>
> Congestion Charge: a scheme was introduced in London whereby vehicles are charged a fee to enter certain parts of the city. This had the immediate effect of reducing the number of cars on the road. Also, the money raised was spent on improving public transport.

> **Tip**
>
> **What are flow, desire and trip lines?**
> These are all used on maps to show movement as either arrows or lines. They can also be used to show the density of the movement.
>
> It is important to note that some textbooks interpret these terms in slightly different ways. Make sure that you have the correct interpretation based on your specification.

#### Tasks

1 Using a photocopy of the flow line map on the next page showing traffic flows into Manchester, add the following data:
  A56    1780
  A6     1480

2 Describe the pattern of traffic flow into Manchester.

3 Suggest one strength and one weakness of this technique.

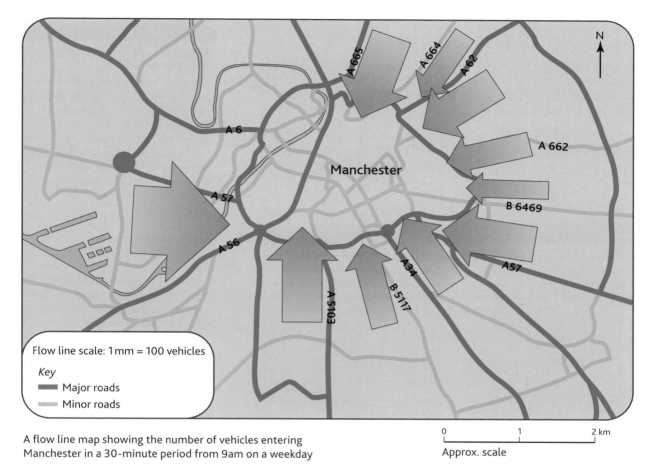

A flow line map showing the number of vehicles entering Manchester in a 30-minute period from 9am on a weekday

0    1    2 km

Approx. scale

# Desire lines

In technical terms, a desire line is a simple concept. Imagine a path constructed around your local park using asphalt. It is unlikely to be the fastest route across the park, i.e. the shortest distance. A desire line would appear as people trample across the vegetation in the shortest distance possible, ignoring the path provided. As others follow, it becomes a wider path. The more people who use it, the deeper and wider it becomes as footpath erosion occurs.

In geography, this idea has been slightly adapted to show how busy the route is between two places. In desire line maps, the line ignores the actual route taken but simply concentrates on the origin, the destination and the number on the route. Desire lines are usually drawn on smaller-scale maps showing movement between regions or even different parts of the world.

The formation of a natural desire line across an urban park. People have forged their own path across this park: this is almost certainly the shortest and fastest route across this part of the park

## Case study

### Investigating migration into the UK

As part of his geography studies, a student conducted a secondary data research task in order to find out the origin of people who permanently settle in the UK. He decided to display this information on a desire line map.

He found that 143,205 people were given grants of settlement in 2006. He chose to display the data in percentage form, focusing on migration by continents of the world.

### Key point

**Grants of settlement versus migrant workers**

This data is about people who have been given the legal right to settle in the UK (grants of settlement). This should not be confused with people who have come to the UK for work purposes (migrant workers), such as those from the European Union (EU). This latter group usually return to their country of origin.

### Take it further

Go to www.migrationinformation.org/datahub/countrydata/data.cfm and create your own desire line map showing migration to the UK from European countries.

### Tasks

1 Complete the table below by using the desire line scale shown on the map below to work out the actual percentages from each continent.

| Continent | % share of UK immigration in 2006 | Estimated numbers from each continent arriving in 2006 |
|---|---|---|
|  |  |  |
|  |  |  |
|  |  |  |
|  |  |  |
|  |  |  |
|  |  |  |

2 143,205 people were given grants of settlement in the UK in 2006. Use this information to estimate numbers arriving from each continent in 2006.

3 Describe the pattern of migration into the UK in 2006.

**Extension**

4 Suggest reasons to explain why the numbers arriving from Asia might be so large, compared with the other continents.

### Key terms

**The sphere of influence:** the maximum distance people are prepared to travel to use that service. A corner shop would have a small sphere of influence (perhaps just a few streets), whereas a supermarket would have a much larger sphere of influence.

**Honeypot site:** a tourist attraction that attracts large numbers of visitors part of the year or all year round. It can be a human or physical location.

A desire line map showing the origin of migrants into the UK from around the world in 2006

# Trip lines

Trip lines are a variation on the desire lines concept. They are used to display information related to the trips or journeys taken by individual people, say from home to work. On a map, trip lines look like spokes on a wheel. By displaying this information on a map, seeing patterns becomes easy. For example, a series of trip lines could be drawn out from a central point (such as a supermarket) to each customer's home. By doing this, it is possible to see the **sphere of influence** of the supermarket. Unlike the other techniques, trip lines do not usually show density and there is no need for a scale for the line.

### Sphere of influence of supermarkets *x* and *y*

*Methodology*

A pair of students wanted to compare the sphere of influence of two supermarkets in their home town, Widnes. They set up a questionnaire that asked for the address of each customer. Before asking the question, they first explained that they were A-level students undertaking some research. Some people chose not to respond, but for those who did, they recorded the addresses in a notebook. Back at school they used an OS 1:50 000 map to show the data using trip lines. Here is the data for one of the supermarkets.

*Case study*

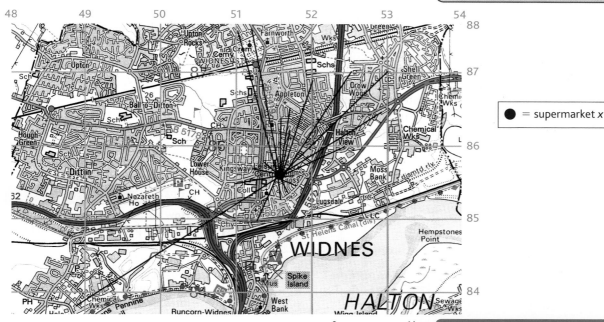

● = supermarket *x*

Trip line map showing origin of customers for supermarket *x*

### Tasks

1 Describe the pattern shown on the map.
2 Estimate the mean distance people travel to this supermarket.
3 Suggest a possible reason why, based on this survey, nobody from Ditton or Upton Rocks appears to use this supermarket.

### Summary

■ Choosing either the wrong scale of map or the wrong scale for your arrows/lines will make your map look very unclear.

■ It is sometimes quite difficult to fit all of your data on the map, so plan what you want to show before you start.

## Tasks

1 Make a copy of the sketch map shown below.

2 Using the formula on page 47 and data provided below for each site, calculate the radius (in millimetres) for each proportional circle for Sites 2–7.

Site 2   1110 ⎫
Site 3   760  ⎪
Site 4   820  ⎬ Calculated estimated volume
Site 5   591  ⎪ in cm³ for each site
Site 6   395  ⎪
Site 7   450  ⎭

3 Now add each correctly proportioned circle to the sketch map. Remember that site 1 has already been calculated on page 47. Choose the positioning of each circle on your map carefully. You should indicate the site to which each circle applies.

4 Add the scale for the proportional circles you have drawn.

### Extension

5 Now describe the pattern on the completed sketch map.

6 Suggest reasons for the pattern you have identified.

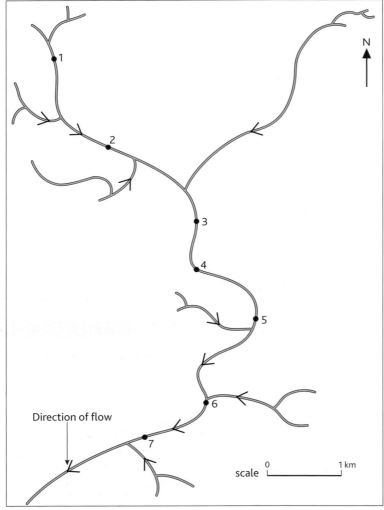

Sketch map for adding proportional circles showing changes in mean bedload volume along river x

# Showing density on maps 1: choropleth map

Choropleth mapping is one of the most widely used techniques in displaying geographical data. It is a simple technique to use and is extremely effective at helping you to observe patterns that would otherwise remain hidden in numerical data. The technique works by using a system of colour, grey scale or line density shading to show how the density of your data changes from place to place.

An important point to remember is that data cannot appear in two categories; for example:

Category 1        0–10
Category 2        10–20

Here, the number 10 appears in 2 categories. Therefore, use the following technique to avoid that problem:

Category 1        0–9.9
Category 2        10–19.9

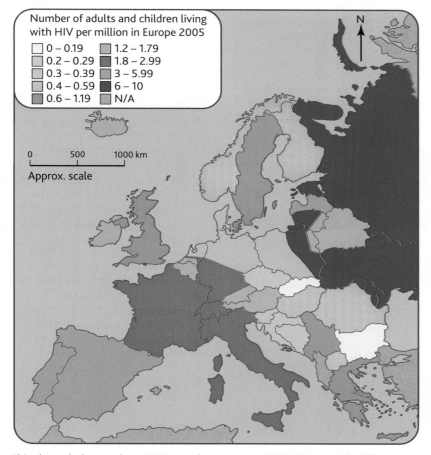

This choropleth map shows HIV rates by country in 2005. Can you identify problems with choropleth maps from this example?

## In this section you will learn:

1  when it is appropriate to use the technique

2  how to calculate the class intervals

3  the difference between shading and cross-hatching

4  some other places where you might use the technique.

## Tip

See histograms (page 67) for more on working out class intervals.

## Key point

### Line density shading

This is also known as cross-hatching. Here, instead of using a colour scheme, a series of increasingly dense lines are drawn.

Let's try an example:

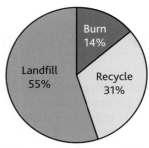

What we do with our rubbish

*Case study*

## Investigating patterns of opposition to a new development

A student received a letter through the post about a meeting in the local town hall. The government was planning a new Energy From Waste (EFW) scheme in his town. They wanted to increase the amount of rubbish that is burned. He decided to see if there was any pattern in terms of who would turn up to the meeting opposing the new incinerator. First, he risk assessed the study with his teacher. At the meeting he told each person that he was a student doing some research into the new incinerator. He first asked if they were in favour or opposition to the new EFW plans. He then asked for the home postcode of each person. Out of the 250 people who attended the meeting, 200 were opposed. All the opponents gave him their postcode. From this data he was able to work out which district they lived in. Once he had the numbers from each district, he was able to turn this figure into percentages.

He decided to display this information on a choropleth map (see below).

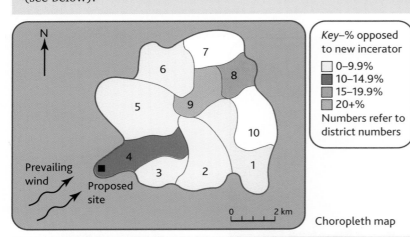

Key–% opposed to new incerator

☐ 0–9.9%
■ 10–14.9%
■ 15–19.9%
☐ 20+%

Numbers refer to district numbers

Choropleth map

### Tasks

1. Make a large copy of the base map shown above.
2. Using the key provided, add the following data for:
   District 7 = 21%
   District 10 = 15%
3. Describe the pattern suggested by this survey.
4. Suggest reasons for the pattern.

**Extension**

5. Use the following percentages to work out the actual numbers who opposed the incinerator from each district:
   District 1 = 1% District 2 = 2% District 3 = 4% District 4 = 16%
   District 5 = 5% District 6 = 4% District 7 = 21%
   District 8 = 21% District 9 = 11% District 10 = 15%
6. Perform a chi-squared test (see page 111) on the data to investigate whether there is a significant difference between districts in the proportion of those attending meetings. Remember to start with a null hypothesis. The chi-squared test would assume that the population of this town was equally distributed between the districts.

### Tip

You could use the choropleth mapping technique to display many sorts of data that relate to different places within an area with recognised boundaries. This includes displaying:

- population densities
- rates of disease
- local authority data, such as unemployment rates.

### Summary

- The technique can only be applied where clear boundaries exist between places, e.g. wards in a city.
- The completed map will hide any variation within each zone. You do not know the actual data at any point on the choropleth map.
- If you use too large a class interval, lots of places are likely to have the same shading, making it difficult to see patterns.
- If you use too small a class interval, you might find that you produce a very large number of classes, making it difficult to see a pattern. This also makes the map quite complicated to produce.
- The completed choropleth map gives the impression that abrupt changes have occurred at each boundary. The reality is probably that change is much more gradual.
- GIS packages such as AEGIS 3 (page 123) are transforming the way maps such as this can be created.

# AS/A2 Showing density on maps 2: isoline maps

Isolines are lines that represent the same value along their whole length. Have a look at contour lines on page 40 or isobars on pages 57–59 as these are both special types of isoline map. Some other uses of isolines are for displaying river depth data and for displaying temperature change in a microclimate study (isotherms).

Unlike choropleth maps, which use formal boundaries such as wards within a city, isolines are themselves the boundaries that mark the change in density.

When using this technique, you need to follow some simple rules carefully:

1 First display your data on your map.

2 Once this is done, you should be able to tell if it is suitable to use the technique.

4 Choose an appropriate scale for your isolines. They should be at equal intervals (e.g. increase in multiples of 10 or 20, depending on what your data range looks like).

5 Start by encircling the highest values.

6 You will need to use a bit of common sense as some places on your map may not 'fit' what you are trying to do; for example, there may be anomalies. It may be better to remove some values and write about why you did so later. Get some help if you are unsure.

Let's now complete the investigation started by the students on page 36.

> **In this section you will learn:**
>
> 1 the difference between an isoline and a choropleth map
>
> 2 when to use an isoline map
>
> 3 how to construct an isoline map
>
> 4 key points about the use of the skill.

> **Tip**
>
> **Deciding whether to use an isoline map**
>
> Look for a steady change in values across the study area, with clear concentrations and values falling evenly and smoothly away.

## Measuring pedestrian flows in a central area

*Case study*

Having drawn their sketch map (page 37), the students were now ready to collect their data.

### Methodology

They agreed on 35 different places to conduct their survey across the central area. They synchronised their watches and all started at the same time. They agreed to count pedestrians travelling in both directions (using a tally chart) for a period of 10 minutes. They tried not to count people twice and they did not count children in school uniform or younger children.

Back at school, they were able to display their findings on the sketch map they had created. First, they added all the pedestrian data to the 35 locations on the map. They were then ready to begin drawing the isoline map.

> **Key point**
>
> **Adding colour**
>
> Once you have drawn your lines, start with the highest density and colour this darkest (perhaps purple). Lines surrounding the peak should become lighter to show density falling away.

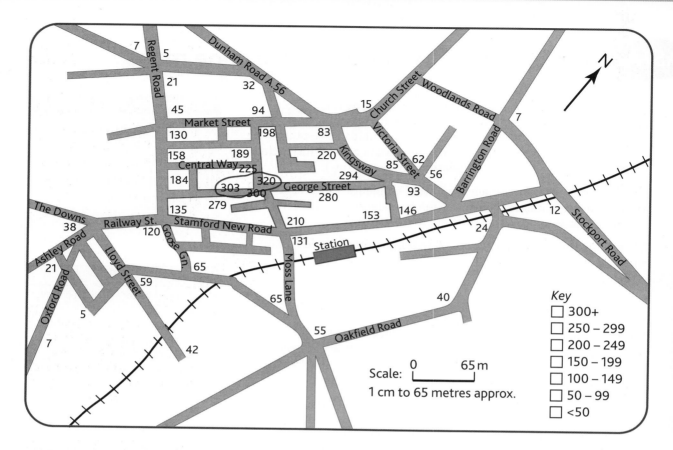

Scale: 0 |———| 65 m
1 cm to 65 metres approx.

**Key**
- ☐ 300+
- ☐ 250 – 299
- ☐ 200 – 249
- ☐ 150 – 199
- ☐ 100 – 149
- ☐ 50 – 99
- ☐ <50

## Tasks

Using a photocopy of the sketch map above:

1 Using the scale provided, draw on each isoline. The first isoline has been drawn for you.

2 Label each isoline in the same way the first one has been done.

3 Choose an appropriate colour scheme and add the colour, starting with highest intensities first.

4 When you have added all the isolines and coloured them in, describe the pattern shown on the map.

### Extension

5 Suggest reasons for the pattern you have identified.

## Summary

- There should be a minimum of 20–30 pieces of data on a typical map.
- The lines should be drawn at equal intervals.
- Lines should only transect (pass through) a point of identical value to the line itself. Isolines are sometimes confused with joining the dots!
- Like choropleth mapping, choosing the correct colour for the key is very important.
- Cross hatching/line density shading is not normally used.
- Unlike choropleth maps, actual values can be identified, rather than just ranges.

## Exam tip

**Describing patterns**

When describing patterns on a graph or map, look for the general trends as well as any anomalies (page 62). A good rule to follow is to 'say what you see'. It is also important to use actual data from your graph or map to support your answer. This could be figures or place names, for example. When the question asks you to describe, you should never give reasons as this is explanation. There will be no credit, even though explaining is usually harder than describing. Command words are considered further in Appendix 2.

## Key point

**Some other uses of isolines**

See contour lines (pages 40) and isobars (pages 57–59), which are both special types of isoline.

Displaying river depth data.

Displaying temperature change in a microclimate study (isotherms).

# AS/A2 Showing density on maps 3: dot maps

Dot maps are a useful way of identifying the **density** of a particular variable such as population. Dot maps also indicate the **distribution** of a particular variable. Unlike a choropleth map, it is possible to estimate the numbers in a particular place, provided each dot is clearly visible.

The map below shows where most of the people are living in Brazil and how concentrated they are in a given location. Clearly, most people live on the east coast with a real concentration in the south-east. Can you work out the exact numbers on this map though?

Dot maps do have their limitations. This map of Brazil seems to suggest that there is nobody living in many areas, when this is clearly not true. This is a scale-related problem. As each dot represents 100,000 people, areas that are more sparse than this would not receive a dot. Dot maps are therefore misleading when there is a large difference in densities between places. In extremely dense places, the dots begin to merge. While this is useful in visually representing high density, accurately counting the population becomes impossible.

## Constructing dot maps

It really depends on the sort of dot map you want to create. For a typical dot map you simply need to place the dot on your map where the feature occurs. For example, if you were mapping arable farms in South East England, you could use a base map and simply place your dots at the centre of each farm. You could also use your dot to represent a particular scaled value, as has been done in the map of Brazil (left). Careful thought has to be given to the scale which should be applied. Here is a simple approach for population dot maps:

1 You first need the total number in a given area, e.g. 20,000 people.

2 Choose a scale which represents this, e.g. 1 dot = 5,000 people.

3 This scale would produce 4 dots.

4 Add these dots at equally spaced intervals to that given area on your base map.

### In this section you will learn:

1 the difference between a dot, isoline and choropleth map

2 when you should use a dot map

3 how to use and interpret a dot map

4 key points about the use of the technique.

### Key terms

**Density:** is about the concentration that exists. Dot maps consider both the distribution and the density.

**Distribution:** is concerned with the pattern or spread of data.

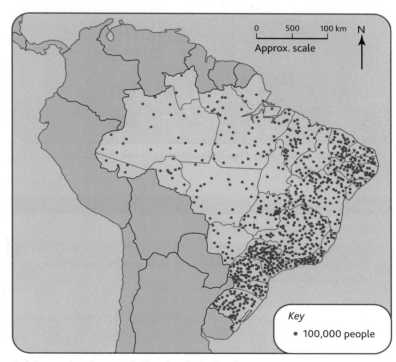

A dot map showing population density in Brazil

## Investigating road transport

As part of her study into road traffic accidents, a student used a dot map to show the number of fatal accidents that had occurred in her home region of the Thames Valley. In the map she produced, each dot represents one fatal accident. The map she created is shown below.

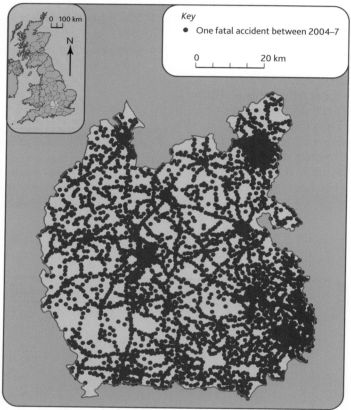

Dot map showing road traffic fatalities between in the Thames Valley, 2004–7

### Tasks

Using the dot map shown above:

1  Describe the distribution of road traffic fatalities in the Thames Valley.

2  State one strength and one weakness of using the dot map technique to display data such as this.

### Extension

3  What other data would it be useful to collect in order to further investigate traffic issues in the study above?

4  How could this data be used to explore relationships?

### Tip

You can use other symbols instead of dots, e.g. squares, triangles and even letters.

### Summary

- Getting the scale right for what each dot represents is crucial.
- Choosing the right scale of base map on which to display the map is also important.
- Where the dot represents a scaled value, low density places are under-represented.
- In high density places, obtaining exact, accurate values is very difficult, particularly when dots merge.
- See located proportional circles for an alternative to dot maps (page 47).

# Constructing and using detailed town centre plans

A detailed town centre plan is just like any other map. The only difference is that the scale is much larger and it is usually only focused upon the central area of a settlement.

We use detailed town centre plans like **Goad plan** or OS MasterMaps when we want to investigate patterns, trends and changes over time in a central area.

## How is the land use classified?

Gathering data for a detailed town centre plan is quite simple. You need to orientate yourself with your map. To do this it is probably easiest to ignore the north arrow and turn the map around so you can identify your location in the map, relative to the buildings around you. This is not the only way of constructing a land-use map, but it is probably the easiest if you have never done it before. Double-check your location by picking out any unusual building (perhaps very large, small or irregularly shaped). Make sure it is in the right place relative to where you are standing.

You are now ready to construct the land-use map. A system known as RICEPOTS classification (see right) is probably the easiest way to get started.

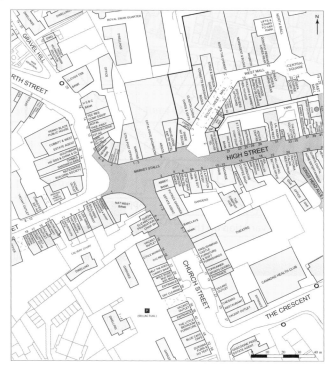

Imagine how long it would take you on your own to classify every building in this central area. For safety reasons you should never work alone anyway. In this size of study area about 6 students working in pairs could complete this task quite easily

### In this section you will learn:

1 why detailed town centre plans are used in geography

2 how to construct an urban land-use map

3 how to use the RICEPOTS system of classification

4 how to interpret a detailed town centre plan.

### Key term

Goad plan: is a detailed town centre map showing individual building outlines and land use at a given point in time.

### Tip

You are likely to use a detailed town centre plan to investigate the clustering of particular services. See nearest neighbour analysis for more on page 114.

### Key point

**RICEPOTS stands for:**

Residential (e.g. houses, flats)

Industrial (e.g. factories)

Commercial (e.g. shops)

Entertainment (e.g. theatre)

Public building (e.g. town hall)

Open space (e.g. park)

Transport (e.g. train station)

Services (e.g. solicitor's office)

### Task

1 Describe some of the likely problems with the RICEPOTS system of classification.

When you look at a detailed synoptic chart, it is likely to contain information on air pressure and fronts (and possibly station circle plots). Station circle plots are generated locally by authorised Met Office sites and they help to build up a picture of how the weather is changing on the ground across the country.

## What is a front?

In simple terms a front occurs when warm and cold air meet. You should already know that the UK is affected by many different types of air mass. So, when a polar maritime **air mass** meets a tropical maritime air mass, a front is formed. Depending on other factors, such as air pressure, these conditions can lead to the formation of a depression. On a weather map, there are specific symbols to show different types of front, three frontal symbols are shown on the right.

Reading weather maps takes practice and there is a good deal of theory to learn before you can really understand what is happening (and why it is happening).

This synoptic chart looks at the main types of front and pressure conditions that you would expect to see on a weather map around the UK. Make sure that you are familiar with the symbols and that you understand the likely conditions at the different fronts. Find out more about surface pressure charts at **www.metoffice.gov.uk**

Some of the more common symbols used to show different types of front are shown below:

**Cold front**
Cold air is arriving, bringing with it cloud and rain. Following this, temperatures fall

**Warm front**
Warm air is arriving, bringing with it cloud and rain. Following this, temperatures rise

**Occluded front (or 'occlusion')**
Here the cold front 'catches up' with the warm front.
The front at the surface will soon disappear

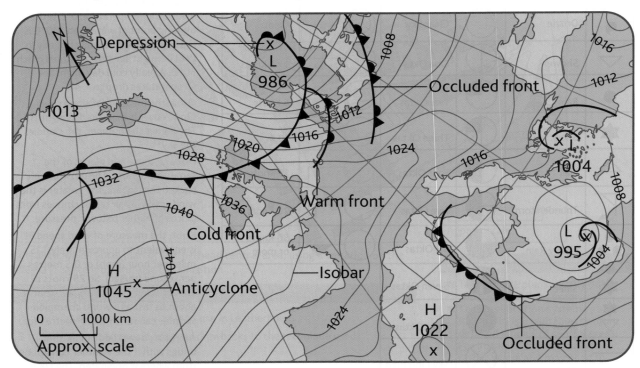

An example of a surface pressure chart showing some typical air pressure readings and frontal situations around the UK and western Europe

> **Tip**
> The section on interpreting digital images (page 119) contains information on how satellite images of the weather can be interpreted.

## Tasks

1 Study the example surface pressure chart shown on the previous page.

2 Describe the likely weather conditions in the UK over the next 24 hours, assuming the low pressure system moves north-east.

3 Study the station circle plot map for the UK below.

4 Create a station circle plot for Fishguard based upon the following information:

| | | | |
|---|---|---|---|
| Wind direction: | SSW | Wind speed: | 20 knots |
| Cloud cover: | 5 Oktas | | |
| Temperature: | 16°C | Present weather: | Rain |

5 Compare the conditions at Fishguard with those in the northwest of Scotland.

### Extension

6 Visit the Met Office website – **www.metoffice.gov.uk** – and create a detailed weather report for the UK for today. Your report should include:

- a current surface pressure image
- a current satellite image
- a detailed description of the different weather conditions experienced across the UK, with appropriate explanation.

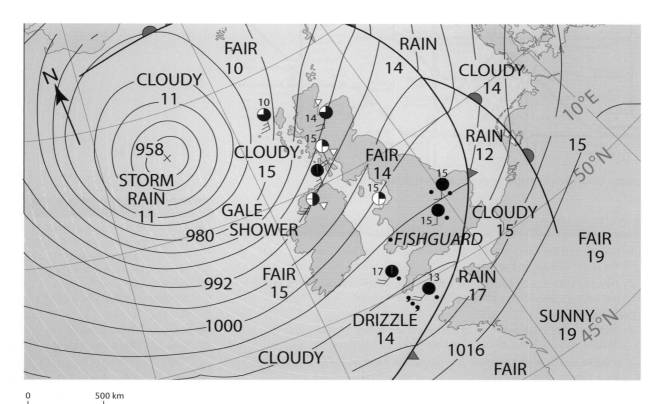

A major low pressure system has formed off the west coast of the British Isles

# Simplifying maps: understanding topological maps

Probably the most well-known topological maps are those of underground rail networks such as the London Underground (see below). There is so much information to show in such a small space that, without bending the rules of map drawing, only the most experienced map users would be able to find their way around. Topological maps bend the rules by ignoring actual distance and direction when they are drawn. Instead, features (such as train stations) are drawn relatively and sequentially to each other. In the case of the underground maps, a simple colour coding system is then added and the train lines are named (or numbered). This then means that even the most inexperienced map users can generally find their way around.

**In this section you will learn:**

1 why topological maps are sometimes used to display data

2 how to find your way around using a topological map.

London Underground map – an example of a topological map

## Tasks

1 Use the London underground map, above, to find your way from Victoria in the south to Angel in the north.

2 Describe the route by naming the stations you would have to change at and the lines you would have to take to get to your destination.

### Extension research

3 Perform an internet search using the term 'Paris Metro'. If you click on 'images' this will bring up lots of maps of the metro system.

4 Imagine you have some friends who are new to Paris; create a route planner for a 'whistle-stop' tour around the city which will take in some of the major sights and attractions that a tourist might be interested in. Name the stations and lines they will need to use, as well as the attractions they can see near each station. Start your route planner at Gare du Nord.

## Summary

■ A topological map is designed to distort actual locations. This means it cannot be used for accurate location referencing. You are never exactly where the map indicates when using a topological map.

■ You are extremely unlikely to have to create a topological map.

■ This section has simply been about showing you what topological maps are and how to use one.

# 3 | Using graphs to transform data

Investigating correlation: scatter graphs

A scatter graph is a type of graph in which the plotted points often appear scattered across the graph paper; hence the name 'scatter graph'. You would use this technique when investigating relationships between two **variables**.

## What is a dependent/independent variable?

If one of your variables is expected to affect a change in the other, the variable affecting the change is referred to as 'independent' and this data is plotted on the *x*-axis (horizontal). The data thought to be affected by the change is referred to as the 'dependent variable' and is plotted on the *y*-axis (vertical). Look at example 1 below.

### Positive correlations

*Example 1: positive correlation using river data*

A straightforward way of understanding **correlation** is to think of some river data. The further down a river you travel, the more water it carries as tributaries (smaller streams) join the river. As distance from source (beginning of the river) increases, so too should discharge. Discharge is the measure used to show the volume of water passing a fixed point, over one second in time. Discharge is usually given in cubic metres per second (cumecs).

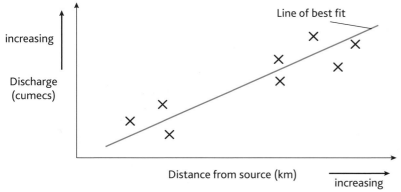

Simple correlation using river data

The relationship above is a positive correlation because an increase in one variable seems to show a corresponding increase in the other. However, this relationship is far from what we would refer to as statistically 'perfect'. A perfect correlation occurs when a unit change in one variable can be matched with a unit change in the other variable. In geographical studies we rarely (if ever) see such relationships. Life on the planet is far more complicated than having one variable directly affecting another in such a neat and tidy way!

### In this section you will learn:

1  the concept of correlation
2  the skill of constructing a scatter graph
3  the strengths and weaknesses of this technique
4  the links to the statistical techniques of Spearman's Rank and Pearson's Product Moment Correlation Coefficients.

### Key terms

**Variable:** the elements of the data you are using (numerical information). This can be anything from pebble sizes on a beach to land rents in a town centre.

**Correlation:** is another word for relationship. It is used to suggest that there might be some sort of link between two pieces of data. As you can see opposite, there appears to be a link between distance from the source and discharge. It appears that when distance from the source increases, discharge increases.

**Line of best fit:** this is usually a straight line drawn to represent the direction and nature of the correlation.

### Tip

**Line of best fit**

The simplest way to draw the line of best fit is to do it by eye and draw a line that broadly represents your pattern. Another approach is to draw a line with an equal number of points on either side.

However, this is what it would look like:

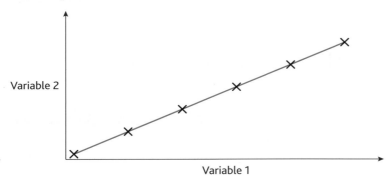

Perfect positive correlation

## Negative correlations

*Example 2: negative correlation using river data*

A negative correlation occurs when an increase in one variable appears to link to a corresponding decrease in the other variable. Look at the graph below, which plots a student's data on pebble size and distance from the source. Our theory would tell us to expect pebble size to get smaller as distance from the source increases. This is because the pebbles will be bashing into each other (attrition) and against the bed and banks during the journey from source to mouth.

> **Tip**
> Negative correlations are easy to understand if you have understood the work so far. If you have not, re-read the section above or get some extra help.

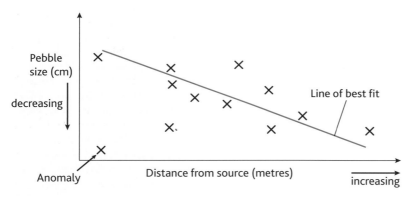

Simple negative correlation using river data

# What is an anomaly?

An anomaly may also be called a 'residual'. Can you see why the cross above has been highlighted as an anomaly? When gathering data in geography we often come across something unexpected. A piece of data that is very different from the rest is an anomaly. In the case above, the pebble is very near the source but very small in comparison with the other findings. Generally, it is good practice to ignore this when drawing the line of best fit, but perhaps refer to it as an anomaly in describing the graph later.

# Investigating pedestrian data

## Methodology

A student gathered some data on pedestrians in relation to distance from a town centre. The pedestrian density was measured by counting how many people passed the person counting for a period of 10 minutes. The sites were determined using street corners at the nearest 50-m intervals (gaps) moving away from the centre of the town. A trundle wheel was used to identify the 50-m intervals along the line of study (transect). A main road was chosen for the transect. The centre of town was taken as the point where land values were at their peak.

Bid-Rent Theory would tell us that town centres are the most accessible parts of the town for large numbers of people. This then attracts the biggest brand stores who need large numbers of customers. These brand stores can pay the high land values. The further away we move from the town centre, the more inaccessible locations become and values decline, attracting smaller stores that are less attractive to shoppers. So, in summary, our theory tells us to expect numbers of pedestrians to fall as we move away from a town centre. Here is the data:

| Distance from centre (metres) | Pedestrian data (numbers per 10-minute interval) |
| --- | --- |
| 0 | 236 |
| 50 | 242 |
| 100 | 154 |
| 150 | 144 |
| 200 | 121 |
| 250 | 46 |
| 300 | 101 |
| 350 | 95 |
| 400 | 43 |
| 450 | 64 |
| 500 | 25 |
| 550 | 12 |
| 600 | 12 |
| 650 | 5 |

Pedestrian data table

## Tip

If you have between 10 and 30 pairs of data, you can conduct a Spearman's Rank Correlation Coefficient test (page 100).

## Summary

- Lots of points can be shown in a relatively small space.
- Patterns can be identified quickly and easily.
- Anomalies can often be identified (see page 62).
- You can add a line of best fit to indicate a possible relationship between two variables.
- You can easily make mistakes when plotting large numbers of points.
- Scatter graphs can only be used to display two variables.
- When you draw on your best fit line, you can suggest a misleading relationship between your two variables, i.e. one that is not actually present.
- The data used in this technique could be further analysed using either a Spearman's Rank or a Pearson's Product Moment Correlation Coefficient (see pages 100 and 103).

## Task

1 Using the data table above and a piece of graph paper:
   a add a title
   b label both axes (be aware of which is more likely to be the dependent/independent variable)
   c choosing an appropriate scale, plot the points to draw a scatter graph
   d add a best fit line
   e describe the correlation.

# Using bar graphs and histograms

A bar graph (or bar chart) in its simplest form is a representation of the number in a set of data, usually in the form of categories. It can be used either to compare different data sets or to compare categories within a set of data.

## Simple bar graphs

Bar graphs can be displayed in a number of different ways, depending upon the data in your set. The x-axis is usually used to display categories and the y-axis to display values in each category. However, there are exceptions to this (see divergent bar graphs on page 65).

Here are some basic points when drawing a bar graph:

1 Always use graph paper where possible.

2 Choose the most appropriate page layout. If you are drawing lots of bars, landscape is probably best. However, it is probably best to use portrait for a large range of data (i.e. high and low values).

3 Choose a simple scale which means the data will easily fit on to the page, e.g. use increments of 5 or 10.

4 Clearly label both axes and add an appropriate title.

5 It is technically correct if the bars are drawn with gaps in between, rather than next to each other. However, it is probably easier to draw the bars next to each other and you are unlikely to be penalised.

6 With some types of bar graph shading or a key will be needed.

### Case study

### Investigating seasonal rainfall variation in the UK

Students were given some average rainfall data for Greenwich in London. They decided to show this on a bar graph.

| Month | Jan | Feb | Mar | Apr | May | Jun | Jul | Aug | Sep | Oct | Nov | Dec |
|---|---|---|---|---|---|---|---|---|---|---|---|---|
| (mm) | 51.9 | 34.0 | 42.0 | 45.2 | 47.2 | 53.0 | 38.3 | 47.3 | 56.9 | 61.5 | 52.3 | 54.0 |

Average rainfall for Greenwich

## Comparative bar graphs

If you have two or more sets of data that you want to compare on the same graph, you would use a comparative bar graph. This type of bar graph might be used for the display of data comparing change over time or for displaying different places within the same axes.

The graph at the top of page 65 shows monthly rainfall averages for different cities around the UK.

### In this section you will learn:

1 how to create a bar graph
2 the different types of bar graph
3 how a bar graph differs from a histogram
4 key points about the use of bar graphs and histograms.

### Key term

**Population structure:** shows the numbers (or percentages) of different age groups and genders that make up a population. Male and female are divided by the y-axis. Provided you have accurate data, structure can be shown for a city, though it is perhaps more likely to be shown for a country. The completed divergent bar graph is often referred to as a 'population pyramid', although many countries' structures no longer resemble a pyramid. Can you work out why this might be the case?

### Tasks

1 Using a piece of graph paper, construct a bar graph to show the average rainfall per month for Greenwich.

2 Describe the variation in average rainfall in Greenwich. In your answer, consider: the total rainfall; the range in average rainfall by month in Greenwich; the wettest season.

### Extension

3 Go to: www.metoffice.gov.uk/climate/uk/averages/19712000/index.html and obtain rainfall figures for Paisley, Scotland.

4 Construct a bar graph for Paisley in the same way as in Task 1.

5 Describe and suggest reasons for the differences in rainfall at these two locations.

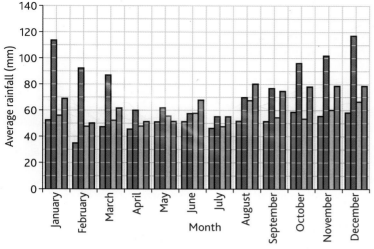

Bar graph showing comparative rainfall data for four English cities

# Divergent bar graphs

A graph with data spread on either side of the *x*-axis is called a divergent bar graph. An example of this might occur when displaying positive and negative values. The same is possible for the *y*-axis. Population pyramids are an example of a divergent bar graph across the *y*-axis.

## Investigating population structure

As part of their studies into population change, a group of students conducted a secondary data search of the internet to gather information on the UK **population structure** in 2001. A partially completed divergent bar graph (often referred to as a 'population pyramid') for the UK is shown below.

**Case study**

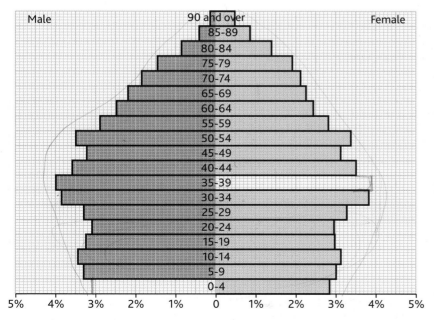

UK population structure, 2001

**Tasks**

1 Describe the general pattern of average rainfall.

2 Outline some of the main differences between the cities.

**Extension**

3 Suggest reasons for the differences by month and city.

**Tip**

For average rainfall, data is collected over a period of at least 30 years.

**Tasks**

1 On a copy of the divergent bar graph on the left, add the following figures to complete the graph for the UK in 2001:
Females aged  35–39  3.9%
Males aged    0–4    3.1%

2 Describe the shape of the graph.

**Extension research**

3 Account for the differences in structure between male and female over the age of 70.

4 Visit the website **www. nationmaster.com/country/ pl/Age_distribution** and print off the population structure for Poland in 2005.

5 Describe the similarities and differences between the Polish and the UK structures.

6 Suggest reasons for the differences.

# Compound bar graphs

Individual bars can also be broken down to show more than one piece of information. This is a compound bar graph. These graphs are also commonly referred to as 'percentage bar graphs'.

**Case study**

## Differences in employment structure

A group of students decided to investigate the **employment structure** of selected countries around the world. They wanted to see if the structure varied in different countries. They used the website **www.cia.gov** to gain access to up-to-date information on all the countries of the world using *The World Factbook*. They were not sure how best to display this information so they decided to use both a compound bar graph and a triangular graph (see page 69). Here is the data they collected:

| Country | Percentage employed in primary industries | Percentage employed in secondary industries | Percentage employed in tertiary industries |
|---|---|---|---|
| Afghanistan | 80 | 10 | 10 |
| Brazil | 20 | 14 | 66 |
| China | 43 | 25 | 32 |
| Ethiopia | 80 | 8 | 12 |
| Germany | 3 | 33 | 64 |
| India | 60 | 12 | 28 |
| Japan | 5 | 28 | 67 |
| Russia | 11 | 29 | 60 |
| UK | 2 | 18 | 80 |

In order to construct this compound bar graph, you need to follow some basic rules:

1 Look at the guidance on how to draw a bar graph on page 64. In this case, it is probably best to draw the graph in landscape.

2 When drawing the vertical scale, remember in this case that all values add up to 100.

3 Remember to divide up the graph carefully, showing percentage employed in each type of industry.

**Key term**

**Employment structure:** a breakdown of the number of people (or percentages) employed in different sectors of the economy of a country. These are usually broken down into primary (raw material extraction), secondary (manufacturing) and tertiary (services).

As the tertiary sector has continued to grow in advanced economies, two new groups of service sectors have emerged. Consequently, the service sector has been further subdivided into quaternary (including research and development, and ICT) and quinary (including fields such as government science and universities).

**Tip**

**Portrait versus landscape**

A4 paper is rectangular. If the longest side is vertical when you hold the page in front of you, you are drawing a graph in portrait format. Similarly, if the longest side is horizontal, you are drawing the graph in landscape format.

Do you know the differences between these types of employment?

## Tasks

1 Use the data provided in the table on page 66 to construct a compound bar graph.

2 Describe the variation between countries in employment structure.

### Extension

3 Suggest reasons for the differences.

4 How useful is this technique in displaying this data?

5 Suggest a way in which the technique could be used more effectively to display this data.

## Constructing a histogram

Another type of bar graph is a histogram. Here, the $x$-axis is used to display classes or groups of data. The $y$-axis is used to show the frequency of occurrence of data within each class. The main difference between this and a normal bar graph is that in a histogram the area of each bar represents the data, whereas in a bar graph it is only the height that represents the data.

Rules for drawing a histogram:

1 Decide on the number of classes (to be shown on the $x$-axis). You can do this using a mathematical method whereby the number of classes is calculated using a scientific calculator:

$5 \times$ log of the total number in the sample. So, if there were 100 pieces of data in the sample you would need to calculate $5 \times (\log 100)$ and then round up or down to the nearest whole number.

Alternatively, just use common sense and decide the number of classes yourself. Get help on the number of classes if you are unsure.

2 The class intervals can be worked out by simply calculating the range of data (i.e. highest to lowest values) and dividing this by the number of classes. This will give equal size class intervals.

3 There must be a scale on both axes. Where class intervals are equal, this is easy.

4 Just as when you construct a choropleth map (page 49), you have to make sure data cannot fall into two classes. For example, 0–10 and 10–20 means the data value 10 can fall into two categories. You can get around this by using class intervals: 0–9.9 and 10–19.9 and so on.

5 The $y$-axis is used to show frequency for each class.

6 Label both axes and add a title.

> **Key point**
>
> Look at the concept of normal distribution (page 92) before continuing. Remember that in a normal distribution, both the median and the mean should have the highest frequencies. The data is skewed if this is not the case.

> **Tip**
>
> The range of data in each class does not need to be equal (though drawing the graph is much easier if they are equal). If the class intervals are not equal, you need to make a scaled adjustment on the width of the bar on the $x$-axis to represent this.

## Displaying long axis measurements

On page 99, the student investigated the long axis measurements of 100 pieces of beach material. Before investigating the standard error of the sample, she first decided to display this information in a histogram. She used a histogram because she was interested in finding out if the data was normally distributed (page 92).

She first had to decide on the number of classes and the range within each class. The formula 5 x (log 100) suggested she should use 10 classes. However, after looking at the range of data in the sample she decided on just eight classes.

To calculate the range for each class she worked out the range for the sample:

Greatest length of long axis = 7.3 cm
Shortest length of long axis = 0.8 cm    Range = 6.5 cm

She then divided this by the number of classes (8). This suggested a class interval of 0.81 cm. For simplicity, she decided to increase this to class intervals of 0.99 cm. Here is the data:

| | Class 1 0–0.99 cm | Class 2 1–1.99 cm | Class 3 2–2.99 cm | Class 4 3–3.99 cm | Class 5 4–4.99 cm | Class 6 5–5.99 cm | Class 7 6–6.99 cm | Class 8 7–7.99 cm |
|---|---|---|---|---|---|---|---|---|
| Frequency | 3 | 8 | 18 | 19 | 27 | 16 | 7 | 2 |

She was now ready to construct the histogram.

# Pictographs

Pictographs are another type of bar graph. Instead of the height of the bar representing values, a picture is used. The pictures have to be of equal size and they usually represent the actual data. For example, a pictograph displaying traffic flow might use pictures of vehicles.

This technique should be used with caution, because where there is no scale (i.e. one unit is equal to one picture), only a limited range of data can be displayed. If the pictures are hand drawn, making sure they are uniform is practically impossible. If a scale is used, drawing a proportion of the picture to represent this is very difficult to interpret accurately.

**World population**

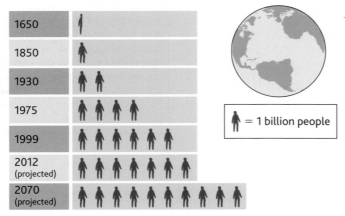

Can you see the limitation of pictographs?

### Summary

■ Bar graphs are versatile graphs with many potential uses.
■ Bar graphs can show positive and negative values.
■ Plotting too many bars makes the graph look cluttered; it becomes difficult to read and easy to miss patterns in the data.
■ If there is a wide range of data to be displayed, the bar graph loses its impact as it becomes difficult to read accurately.
■ Histograms become more complicated to construct when uneven class intervals are used.
■ When using histograms, too many or too few classes can hide important patterns in the data.

Using triangular graphs

Constructing a triangular graph may look daunting and can appear confusing, but it is a really easy skill to learn. Interpreting the technique is a little more difficult, but by following simple guidelines, you can easily make sense of data displayed on a triangular graph. You need to understand that triangular graphs display information on three axes and not two axes as you would normally expect with graphs.

You can only use a triangular graph if you have three variables that add up to 100. For example if employment structure is divided into primary, secondary and tertiary (i.e. three variables). These figures are usually given as percentages of the working population (i.e. represented out of 100). Before attempting the tasks, look at the example which shows how to identify values on a triangular graph shown on page 70.

**In this section you will learn:**

1 how to use a triangular graph
2 how to interpret data displayed on a triangular graph
3 other places where you could apply the technique
4 key points about the use of the technique.

**Tasks**

1 Using a piece of triangular graph paper:
   i Plot the following data, which shows employment structure by selected country:

| Country | Percentage employed in primary industries | Percentage employed in secondary industries | Percentage employed in tertiary industries |
|---|---|---|---|
| Afghanistan | 80 | 10 | 10 |
| Brazil | 20 | 14 | 66 |
| China | 43 | 25 | 32 |
| Ethiopia | 80 | 8 | 12 |
| Germany | 3 | 33 | 64 |
| India | 60 | 12 | 28 |
| Japan | 5 | 28 | 67 |
| Russia | 11 | 29 | 60 |
| UK | 2 | 18 | 80 |

   ii Label each axis and add a title.
2 Describe the pattern shown.

**Extension**

3 What are the strengths and weaknesses of using a triangular graph to display data compared with a compound bar graph (page 66)?

**Tip**

Try the internet to download your own graph paper – http://geographyfieldwork.com/images/Graphs/triangular.gif or see Appendix 6 for a sample piece of triangular graph paper, which you can photocopy.

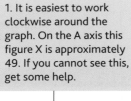

1. It is easiest to work clockwise around the graph. On the A axis this figure X is approximately 49. If you cannot see this, get some help.

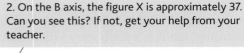

2. On the B axis, the figure X is approximately 37. Can you see this? If not, get your help from your teacher.

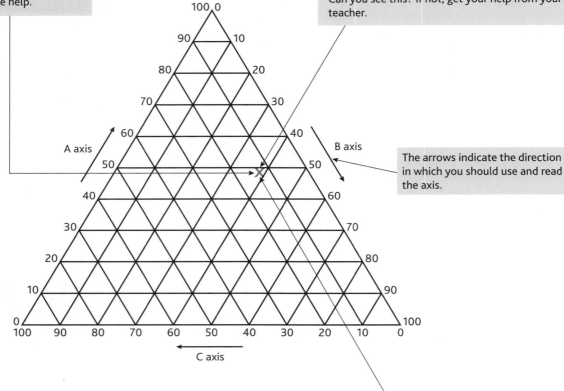

The arrows indicate the direction in which you should use and read the axis.

3. The C axis must therefore be the 100 − (A + B) i.e. 100 − (49 + 37) = 14.
Can you see that X is 14 on the C axis? If you can, you now know how to read data on a triangular graph.

Plotting data on triangular graph takes a little practice

It is only when you plot many pieces of data that patterns begin to become apparent.

### Summary

- The triangular graph is useful in showing patterns of clustering between the three different variables.
- Triangular graphs only work with a very limited range of data: where there are three variables in percentage form.

# Using line graphs

A line graph is a simple but highly effective way of showing **continuous data**. The *x*-axis is usually used to show change over time. The *y*-axis is used to display data such as population or temperature. Getting the right scale is very important. If there are very large extremes within the values, the line graph may need to be plotted on a logarithmic scale (page 69).

Line graphs are useful because they can suggest trends over time and can be used to estimate future patterns based on present trends. Line graphs can also be used to 'fill in gaps' where no data exists. For example, in 1941 there was no census in the UK. However, by using the 1931 and 1951 data it is possible to estimate the 1941 population. Care must be taken when comparing different graphs, as useful comparison can only be made if a similar scale has been used.

Like all other graphs, pay close attention to detail when plotting figures; always label axes and give a suitable title. Remember also to decide whether to draw the graph paper in portrait or landscape format (see page 66). When joining the points on a line graph, always use a smooth freehand style rather than a ruler. This will show gradual change as opposed to sudden change (which may be inferred if you use a ruler to join the points).

## Combining line and bar graphs 1 – the climate graph

A climate graph uses average data for rainfall and temperature over a period of at least 30 years to show how these two variables change over a year. The 12 months of the year are plotted on the *x*-axis. One vertical *y*-axis is used to show rainfall (usually the left axis) and a second *y*-axis (usually on the right of the graph) is used to show temperature change. Bars are always used to display rainfall and a line is always used to show temperature change. See the partially completed example below.

### In this section you will learn:

1 how to complete a climate graph
2 the difference between simple, compound, comparative and divergent line graphs
3 how to construct and interpret a storm hydrograph
4 how to construct a Lorenz curve
5 the difference between a long section and a cross section
6 how to draw a sketch cross section using an OS map
7 key points about the use of line graphs.

### Key term

**Continuous data:** includes variables such as temperature or time. These variables can be broken down into a theoretically infinite number. This data can be shown on a line graph. Discrete data is the opposite of continuous. This data has an actual number that can be counted. For example, counting cars in a car park would be discrete data. This data is best shown on a bar graph.

### Completing a climate graph

A group of students were asked to complete a climate graph for Greenwich, UK. First, they obtained the average rainfall data (page 64) for Greenwich. They were also then ready to add average temperature data to complete the climate graph.

*Case study*

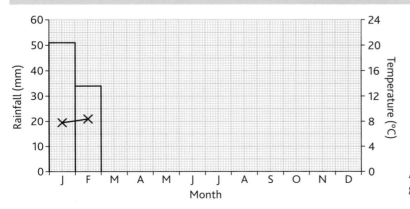

A partially completed climate graph for Greenwich, UK

## Tasks

1  Make a large copy of the partially completed climate graph on page 71.

2  Use the data provided below to complete the climate graph for Greenwich. January and February have already been plotted.

| Month | Jan | Feb | Mar | Apr | May | Jun | Jul | Aug | Sep | Oct | Nov | Dec |
|---|---|---|---|---|---|---|---|---|---|---|---|---|
| Temp (°C) | 7.9 | 8.2 | 10.9 | 13.3 | 17.2 | 20.2 | 22.8 | 22.6 | 19.3 | 15.2 | 10.9 | 8.8 |
| Rainfall (mm) | 51.9 | 34.0 | 42.0 | 45.2 | 47.2 | 53.0 | 38.3 | 47.3 | 56.9 | 61.5 | 52.3 | 54.0 |

3  Describe the climate of Greenwich.

### Extension

4  Go to www.metoffice.gov.uk/ climate/uk/averages/19712000/ sites/paisley.html and obtain average rainfall and temperature data for Paisley, Scotland.

5  Construct a climate graph for Paisley.

6  Describe and explain the main differences between the climate of Greenwich and that of Paisley.

# Combining bar and line graphs 2: the storm hydrograph

The storm hydrograph is a specialised graph that, like a climate graph, uses a combination of bars and lines. Essentially the storm hydrograph shows the impact of rainfall in a drainage basin upon the discharge at a specified point in the river channel. Rainfall is measured in millimetres and discharge is measured in cubic metres per second ($cm^3/s$ or cumecs). Here is how to create a storm hydrograph:

1  Use the *x*-axis to indicate progression of time, usually measured in hours.

2  Use the *y*-axis to show two pieces of information: rainfall, measured in millimetres; and changing discharge within the river (in cumecs).

3  Plot bars first to show the changing rainfall during a storm.

4  Plot changes to the discharge as crosses on the graph and join with a smooth curve.

5  Add a suitable title and label axes appropriately.

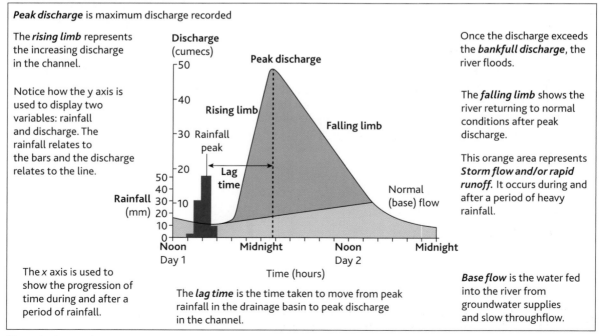

**Peak discharge** is maximum discharge recorded

The **rising limb** represents the increasing discharge in the channel.

Notice how the y axis is used to display two variables: rainfall and discharge. The rainfall relates to the bars and the discharge relates to the line.

The x axis is used to show the progression of time during and after a period of rainfall.

The **lag time** is the time taken to move from peak rainfall in the drainage basin to peak discharge in the channel.

Once the discharge exceeds the **bankfull discharge**, the river floods.

The **falling limb** shows the river returning to normal conditions after peak discharge.

This orange area represents **Storm flow and/or rapid runoff.** It occurs during and after a period of heavy rainfall.

**Base flow** is the water fed into the river from groundwater supplies and slow throughflow.

Storm hydrograph

**Tasks**

Referring to the storm hydrograph on page 72:

1 How much rain fell during the storm?
2 Calculate the lag time.
3 Describe the shape of the hydrograph.

**Extension**
4 Suggest why the rising limb is steeper than the falling limb.
5 Explain the impact on the hydrograph of:
   • afforestation in the drainage basin
   • development such as road building in the drainage basin.

# What is the difference between compound and comparative line graphs?

As with bar graphs, there are many different types of line graph. A simple line graph displays just one set of data, such as temperature on a climate graph. A compound line graph shows more than one piece of information. For example, it is possible to look at total world population growth and break this down into More Economically Developed Countries (MEDCs) and Less Economically Developed Countries (LEDCs). This sort of graph would have three different pieces of information on it.

**Key point**

**MEDC** or **LEDC**?
It is becoming increasingly difficult to clearly define whether a country is 'more' or 'less' economically developed. Countries such as China and India would traditionally be referred to as LEDCs, but this is changing as these countries experience rapid economic development. The term 'Newly Industrialised Countries' (NICs) is sometimes applied to such countries. Visit http://en.wikipedia.org/wiki/Developed_country to find out more.

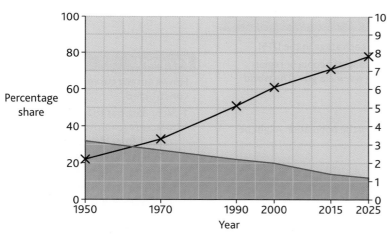

Can you see what is predicted to happen as total world population continues to grow over time?

**Tasks**

1 Which sort of line graph is shown in above. Give reasons for your answer.
2 Identify the main changes shown on the graph.

**Extension**
3 Suggest reasons for the changes identified.

A comparative line graph is different to a compound line graph in that different sets of data can be displayed on the same graph to allow comparisons to be made.

# The Demographic Transition Model (DTM)

The DTM is a specialised comparative line graph which looks at how changing birth and death rates impact upon the total population.

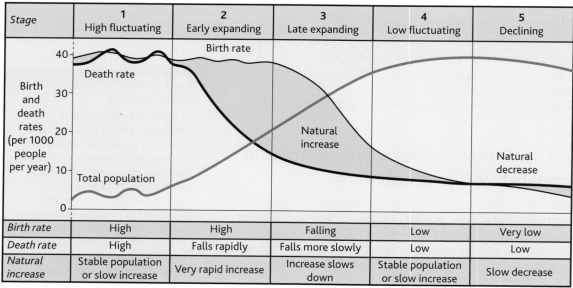

| Stage | 1 High fluctuating | 2 Early expanding | 3 Late expanding | 4 Low fluctuating | 5 Declining |
|---|---|---|---|---|---|
| Birth rate | High | High | Falling | Low | Very low |
| Death rate | High | Falls rapidly | Falls more slowly | Low | Low |
| Natural increase | Stable population or slow increase | Very rapid increase | Increase slows down | Stable population or slow increase | Slow decrease |

The Demographic Transition Model (DTM)

## Tasks

1  Using actual data from the graph above describe each stage in the DTM.
2  Using your own knowledge, suggest a country, group of people or region which might 'fit' into each stage of the DTM. Use www.cia.gov (*The World Factbook*) to research this further.

**Extension**

3  Draw a sketch population pyramid to suggest how the population might be structured in each stage.

# The Lorenz curve

This fairly complicated technique uses the concept of cumulative frequency. It can be used to measure the degree of concentration within a distribution.

The Lorenz curve works by comparing your data with that of a theoretical even distribution. The greater the distance between your plotted data and the even distribution line, the greater the degree of concentration. The easiest way to understand Lorenz curves is to try an example.

## Take it further

You can also use the location quotient technique to investigate concentrations further (see page 117).

## Key points

**How a Lorenz curve is constructed**

1  Your data has to be in percentage form and has to be ranked from highest to lowest.

2  Label both the *x*-axis and *y*-axis 'cumulative frequency'; scale for each axis should be the same, i.e. 0–100%. It is probably easiest to use increments of 10.

3  Draw the even distribution line as a 45° diagonal straight line from 0,0 to 100,100

4  Your figures need to be turned into cumulative percentages. This simply means adding the second figure to the first. Then add the third figure to this new total, and so on. Your total should be 100. If it is not, something has gone wrong and you should start again.

5  Plot the points.

## Investigating social class at football matches

A pair of A-level students were interested in finding out whether a London-based Premiership football club had a different social class of spectator compared with a Championship club (lower division).

They used a simple questionnaire to ask the occupation of the spectator.

They then used the census to obtain national data on the social grade of the population as a whole. Here is the data they collected:

| Social grade | National % | Premiership club % | Championship club % |
|---|---|---|---|
| A/B | 25.2 | 35.4 | 19.4 |
| C1 | 29.8 | 38.3 | 31.2 |
| C2 | 18.3 | 12.1 | 20.4 |
| D | 20.5 | 8.1 | 25.9 |
| E | 6.2 | 6.1 | 3.1 |

To construct the Lorenz curve for the Premiership club, you first need to rank order the data and then turn the percentages into cumulative percentages.

| Social grade | Premiership club % (by rank and cumulative) | National % (cumulative) |
|---|---|---|
| C1 | 38.3 | 29.8 |
| A/B | 73.7 | 55.0 |
| C2 | 85.8 | 73.3 |
| D | 93.9 | 93.8 |
| E | 100.0 | 100.0 |

Can you see how these figures have been created? If not, go back and read the explanation again, or get some help.

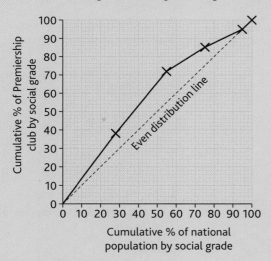

A Lorenz curve showing the cumulative social grade of Premiership spectators against national data

A glance at the above Lorenz curve shows that the Premiership club has a concentration of C1 and AB social groups. This means that in this particular study, there is a disproportionate amount of higher social classes attending the match compared with national averages.

### Tip

The students used a systematic sampling method to obtain 100 responses. From this they were able to identify social grade based on the following categories:
A/B – Upper middle and middle class
C1 – Lower middle
C2 – Skilled working class
D – Unskilled working class
E – Unemployed

### Tasks

1 Using a piece of graph paper, construct a Lorenz curve using the data for the Championship club.

2 Describe the graph.

3 How is the graph different from the Lorenz curve for the Premiership club?

**Extension**

4 Suggest reasons for the differences.

# The divergent line graph

A divergent line graph is much the same as a divergent bar graph (page 65). If data is displayed either side of the *x* or *y*-axis, it would be referred to as 'divergent'.

Case study

## Displaying microclimate data

A group of A-level students were investigating the urban heat island effect (see right). During an autumn evening, they took temperature readings at 100-m intervals along a transect (see page 85) stretching from the town centre to the edge of town. Transect points were agreed in advance. They wanted to see whether any changes in temperature occurred. They synchronised their watches and all took their readings at the same time using digital thermometers. They each took three readings and then took the mean (see page 91) of the three readings. They decided to display their data on a divergent line graph.

First, they collected all readings. They then took the mean of the overall data set. This mean became the starting point for the *x*-axis. Finally, they recorded the difference from the mean at each point along the transect. They were ready to draw the graph.

Urban heat island

Cities are generally warmer than the surrounding countryside for many reasons. One reason is that little vegetation or evaporation causes cities to remain warmer than the surrounding countryside. Can you think of any more?

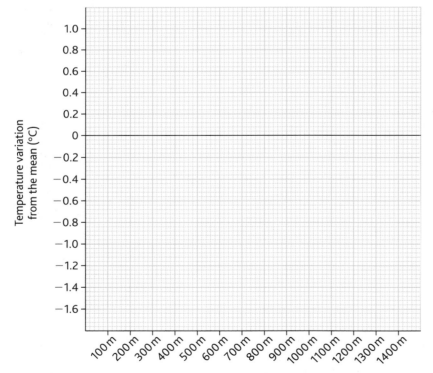

A divergent line graph for showing temperature change with distance from town centre x

## Tasks

1  On a piece of graph paper, make a large copy of the axes of the divergent line graph shown opposite.

2  Calculate the mean for the data set below.

3  Work out the difference from the mean of each value in the data set.

| 100 m | 200 m | 300 m | 400 m | 500 m |
|---|---|---|---|---|
| 14.3 (°C) | 14.5 | 14.0 | 14.1 | 13.8 |

| 600 m | 700 m | 800 m | 900 m | 1000 m |
|---|---|---|---|---|
| 14.0 | 13.5 | 13.8 | 13.2 | 13.0 |

| 1100 m | 1200 m | 1300 m | 1400 m | |
|---|---|---|---|---|
| 14.0 | 12.9 | 12.6 | 12.0 | |

4  Plot this data on your graph.

5  Describe the pattern shown on the graph.

### Extension

6  Suggest reasons for the changes shown.

7  Offer suggestions as to how the data collection methodology could be improved.

# Cross section and long section sketching

These graphs are used specifically to show changing height on an OS map. Cross sections allow you to investigate a particular section of a river and draw the shape of the river or valley to scale at that point. It can then be used later to compare other sections of the river valley. This technique can also be used to investigate valley features created by ice. Long sections use the same principle. However, instead of measuring across the channel perpendicular to the river, a long section looks at changes along the long profile. This is useful in identifying changes in the shape of the long profile and can be used to identify specific landforms such as waterfalls (or knickpoints) caused by changes in the river's base level.

A knickpoint can be formed when a river's base level changes (for instance after sea level change)

A river often affects the shape of the land through the various processes of erosion and transportation. In the hills, narrow steep V-shaped valleys can be formed, while in the lowlands wider valleys often dominate the landscape.

## Drawing cross sections and long sections

1 Choose an appropriate scale of map. 1:50 000 lacks detail and it is very hard to draw an accurate cross section where contours are too closely packed. A 1:25 000 is therefore recommended. However, a 1:50 000 map may be useful for drawing a long section, especially where there is a gentle change in gradient.

2 Focus in on your study area. Your teacher may direct you to this. If you are drawing a cross section on your own, it is a good idea to plot your cross section in such a way that the river lies in the middle, with equal distances up the valley either side. However, this depends on the shape of the valley.

3 Quite often the valley is not of equal slope angle either side of the channel. In this case, choose a common-sense place to end the cross section such as the top of the valley on either side.

4 Draw a thin pencil line on the map (provided it is your own) and mark the start and finish of the line with distinguishing letters (such as X and Y).

5 Use a piece of paper to mark each change in height across the valley. Record the height at each point where the contour line intersects the paper. Remember to mark on the location of the river.

6 On a piece of graph paper, draw a suitably scaled graph to represent the changing height across the valley. The $x$-axis is used to record the points where the changes in height occur across the valley. The $y$-axis is used to record the actual heights at each point where the change occurs.

7 Label the axes and add an appropriate title. Don't forget to mark on the river.

> **Tip**
>
> You are likely to come across these techniques in river studies, glaciation studies or possibly coastal studies. If you have not done so yet, you should now read about contour lines on OS maps (page 40).

> **Tip**
>
> **What scale?**
>
> The horizontal scale should be the same as used on the map.
>
> The vertical scale may need to be exaggerated. You may wish to use a scale of 1 cm = 100 m. Can you work out why?

## Case study

### Comparing channel and valley characteristics

As part of his preparatory work for a river study, a student was interested in seeing how the North Teign River valley changed downstream. The North Teign River lies in Dartmoor National Park. He decided to draw a cross section at a point along the river's course. The map he used is shown in Appendix 4.

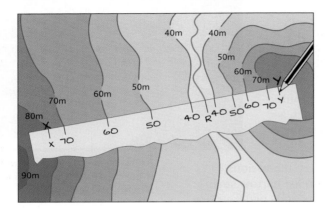

- Once you have decided the two points for your cross section, take a piece of paper and line it up between the points, carefully marking the start and end.
- Mark each change height with a pencil mark on your piece of paper, remembering to label each point with the corresponding height on the map.

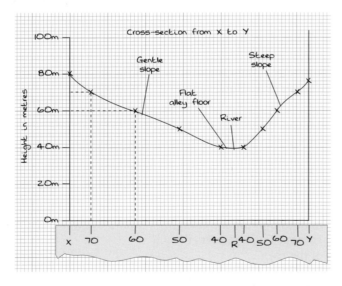

- Draw a suitably scaled *y*-axis on a piece of graph paper. Remember that you may need to exaggerate the scale.
- Using the same piece of paper as above, mark the position of changes in height along the *x*-axis of your graph paper.
- Now use the *y*-axis to mark on the changes in height.
- Finally join the points and add appropriate annotation.

### Tasks

1 Draw a sketch cross section between points A and B as shown on the map in Appendix 4.

2 Describe the valley and channel cross section you have drawn.

**Extension**

3 Suggest reasons for the shape of the cross section you have drawn.

4 Outline the changes which are likely to occur to the channel and valley further down the stream.

5 Suggest reasons for the changes.

### Summary

■ There are many different types of line graph. Choosing the type of graph that will best represent your data is very important.

■ Line graphs are especially useful for displaying continuous data.

■ If using more than one line on a line graph, you should include a key.

■ Too many lines on a line graph can be confusing, especially on a compound or comparative line graph.

# Dispersion diagrams, and box and whisker plots

Dispersion diagrams allow you to investigate visually the spread of a set of data. By plotting all of the data set on a vertical axis, the range within the data set becomes visually apparent. It is also possible to identify any clustering within the data set. Box and whisker plots can be added to analyse the data further. These work by removing the extreme values within the data and focusing only on the interquartile range (see page 94). Dispersion diagrams of more than one data set can be constructed. Provided the same scale has been used, it is possible to compare the dispersions of different data sets to see how range and clustering vary.

## Drawing a dispersion diagram

1  The graph should have a very narrow base and a long vertical axis covering all of the values in the data set. You do not have to start at zero for the y-axis, but if you want to compare dispersions, you have to use the same scale.

2  The x-axis should be labelled with the name of the data set. The y-axis should cover the full range of values for the data set and be labelled accordingly.

3  Plot all values. Where values are equal, plot these adjacent to each other. It may also be useful to write the actual numbers on the graph. This depends on the accuracy that is required and amount of data you are displaying.

4  Add an appropriate title.

### In this section you will learn:

1  how to construct a dispersion diagram

2  how to use box and whisker plots on a dispersion diagram

3  the link between dispersion diagrams, range, interquartile range and median.

### Key terms

Prevailing wind: the direction the wind blows on most days. In the UK, this is a south-westerly wind. This brings moist air to the west of the UK for most of the year.

Rain shadow: is an expression linked to relief rainfall. Once rain falls in the hills, the air loses moisture. As it sinks over the other side of the hills it warms and evaporation occurs. Rainfall totals are lower in rain shadow areas and there is less chance of rainfall.

### Comparing dispersion of rainfall

Case study

As part of an introduction to regional climate variations in the UK, a student was asked to gather some data on the dispersion of rainfall in different parts of the UK. He decided to compare the dispersion of rainfall in the north-west of England with that in the south-east. He knew that the north-west receives **prevailing winds** and that the south-east is in the **rain shadow**. He decided to display this information on a dispersion diagram. He found the data from the Met Office website:

www.metoffice.gov.uk/climate/uk/averages/19611990/areal/england_nw_&_wales_n.html

### Tip

Have a look at median, range and interquartile range, pages 91–95.

| North-west | Month | Jan | Feb | Mar | Apr | May | Jun | Jul | Aug | Sep | Oct | Nov | Dec |
|---|---|---|---|---|---|---|---|---|---|---|---|---|---|
| Rainfall | (mm) | 127.5 | 87.7 | 102.2 | 76.7 | 79.4 | 82.6 | 86.2 | 110.1 | 118.2 | 132.5 | 136.1 | 138.1 |

| South-east | Month | Jan | Feb | Mar | Apr | May | Jun | Jul | Aug | Sep | Oct | Nov | Dec |
|---|---|---|---|---|---|---|---|---|---|---|---|---|---|
| Rainfall | (mm) | 77.7 | 52.6 | 62.2 | 51.9 | 55.9 | 55.0 | 49.0 | 59.2 | 66.8 | 75.0 | 78.7 | 80.6 |

The student then decided to add box and whisker plots. He did this because he wanted to compare the interquartile range with the full set of values. The diagram is shown below.

## Adding box and whisker plots

Follow these simple instructions to create a box and whisker plot for your data.

1  Work out the median (page 92) for the data set. Draw the median line through the median value as a horizontal line (about 1–2 cm across).
2  Calculate the upper quartile and lower quartile (see page 94) for the data set. Draw the upper and lower quartile on the diagram at the appropriate points.
3  Complete the 'box' by drawing on the appropriate vertical lines.
4  Add a thin vertical line through to the highest and lowest values and complete the 'whiskers' by drawing on a horizontal line through the highest and lowest values.

**A dispersion diagram showing variation/dispersion of average rainfall (1961-90) for NW England**

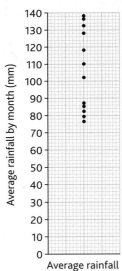

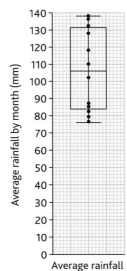

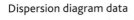
Dispersion diagram data

The data with the box and whisker plots added

### Tasks

1  Using the data provided for the south-east, construct a dispersion diagram showing average rainfall by month.
2  Work out the median, upper quartile and lower quartile.
3  Add these to the sketch as a box.
4  Now add the whiskers.
5  Describe the main differences between the two diagrams.

**Extension**
6  Suggest reasons for the differences.

### Summary

- This technique is a useful visual representation of the dispersion in a data set.
- Dispersion diagrams usually display only one data set at a time.
- If you want to compare many different data sets, it is quite a time-consuming technique.
- You can use this technique in conjunction with other statistical techniques such as standard deviation (page 96).

**Pie charts and proportional divided circles**

Geographers frequently use pie charts to display data. Various components of a whole set of data can be broken down and displayed as a series of segments. The segments are proportional to each other within the pie chart. A proportional divided circle does much the same, but the area of the circle is proportional to the overall values in the data set. These are mainly created for the purposes of comparing data sets.

## Creating a pie chart

1 Each category within the data set has to be turned into a percentage of the set of data. To do this, divide the segment value by the overall value and then multiply by 100.

2 Each category then has to be turned into a number of degrees. Since there are 360 degrees in a circle, you need to multiply your percentage by 3.6 to turn it into a number of degrees.

3 Draw a suitably sized circle. There is no set rule for this. If you are showing a lot of categories (5–6) you should try to draw a fairly large circle, perhaps with a radius of 6 cm. If you were comparing two pie charts, a radius of 4 cm will allow you to place your circles next to each other on an A4 piece of paper.

4 First, draw a line from the centre of the circle to the top. You would normally choose to place each segment in order of size from largest to smallest, working in a clockwise direction. This may make it easier to compare later. You could also add the percentage for each segment for ease of comparison.

5 Use a protractor to construct the segments.

6 Add a key for each segment.

**Tip**

Have a look at located proportional circles on page 47. Here, the symbols are shown on maps.

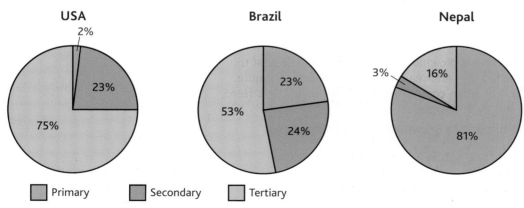

Look at how easy it is to compare the employment structures of these countries using pie charts. Do they tell us how many people are employed in each industry sector in each country?

## Displaying questionnaire responses

As part of her leisure studies, a student was investigating the popularity of Lake Windermere, a key attraction in the Lake District. She devised a questionnaire with her teacher. A risk assessment was first carried out. This is the question she asked:

*'How many times have you visited this location?'*

After she got 50 responses she stopped the questionnaire. She analysed the responses and decided to put them into a series of categories:

| Number of prior visits | 0 | 1–5 | 6–10 | 11–15 | More than 15 |
|---|---|---|---|---|---|
| Number of responses | 4 | 15 | 12 | 8 | 11 |

28.8°  108°  86.4°  57.6°  79.2°

Lake Windermere, a key attraction in the Lake District

### Tasks

Using the Lake Windermere data above:

1 Draw a pie chart with a radius of 5 cm.

2 Convert the number of responses in each category into percentages.

3 Convert each percentage into a number of degrees.

4 Put the categories in rank order and draw the pie chart.

5 Add:
   the percentages for each category
   a key for each category
   a suitable title.

6 Describe the chart.

### Extension

7 Perform a chi-squared test (see page 111) to see if there is a statistically significant difference between the number of responses in each category.

## Creating a proportional divided circle

The creation of each segment of the proportional circle is done in exactly the same way as for a pie chart. The key difference is that the total values are converted into an area for the circle using a formula. This is particularly useful when comparing different sets of data. Here is how to create the proportional circle:

1 You first need a total number ($v$) which you are planning to represent.

2 Substitute the total number ($v$) into the formula:

$r = \sqrt{\left(\dfrac{v}{\pi}\right)}$, where $\pi = 3.142$.

3 This calculation will give the radius ($r$) of the circle.

## Displaying river bedload data

A student wished to display the data he collected for river bedload during a field visit (see also pages 47–48). He collected his data at three different points along the course of the river by taking samples of the bedload. He decided to use proportional circles to display this information.

### Methodology

At each site he took the 3 axis measurements of 20 pieces of bedload and then recorded the data. He also used Powers' Scale of Roundness (see below) to decide the roundness of each piece of bedload. Back at school he was able to work out the average estimated volume (in cm³) for the bedload at each site. He used this data to draw his proportional circles.

This is how he worked out:
- The proportional circle size for Site 1.
- The segments displaying Powers' Scale of Roundness for Site 1.

### Site 1

Formula for area:  $r = \sqrt{\left(\dfrac{v}{\pi}\right)}$

Remember that $\pi = 3.142$

Also remember that in this case the final radius calculation has to be converted to millimetres (or it will not fit on the page).

$v$ = volume of the bedload.

980 cm³ is the calculated volume.

$$r = \sqrt{\dfrac{980}{3.142}}$$

$$r = \sqrt{311.90}$$

$$r = 17.7 \text{ (this means the radius of the proportional circle for Site 1 is 17.7 mm)}$$

> **Formula**
>
> Area: $r = \sqrt{\dfrac{v}{p}}$
>
> Remember that $\pi = 3.142$

### Roundness

|  | Very angular | Angular | Sub-angular | Sub-rounded | Rounded | Well rounded |
|---|---|---|---|---|---|---|
| Number | 5 | 9 | 3 | 2 | 1 | 0 |
| Conversion to degrees | 90 | 162 | 54 | 36 | 18 | 0 |

Very angular    Angular    Sub-angular  Sub-rounded  Rounded  Well rounded

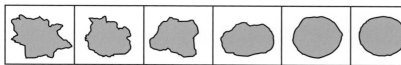

The Powers' Scale of Roundness

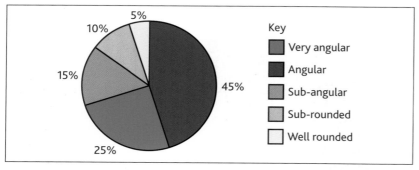

A proportional divided circle showing estimated bedload volume and Powers' Scale of Roundness for Site 1

1  Using the example above and data provided below for each site; calculate the radius for each proportional circle for Sites 2 and 3.

|  | Estimated volume (cm³) |
|---|---|
| Site 2 | 1110 |
| Site 3 | 760 |

2  Construct two proportional circles.

3  Using the data below, construct the segments of these circles.

|  | Very angular | Angular | Sub-angular | Sub-rounded | Rounded | Very rounded |
|---|---|---|---|---|---|---|
| Site 2 | 1 | 3 | 8 | 3 | 3 | 2 |
| Site 3 | 0 | 1 | 4 | 5 | 5 | 5 |

4  Describe the main differences between the three proportional divided circles.

**Extension**

5  Suggest reasons for the differences you have identified.

## Summary

■ Pie charts give a good visual representation of the data.

■ They are best used when there is a wide range of values within the categories. It is difficult to differentiate between categories with similar values.

■ Too many categories (more than six) make a pie chart difficult to interpret.

■ Too few segments (fewer than three) make a pie chart look rather simplistic. Consider using another technique such as a bar graph (page 64).

■ Using an appropriate scale is the biggest challenge with proportional divided circles. Practise with different scales to get the right balance for the display of data. The easiest way to do this is by using millimetres or centimetres, depending on your calculated value for the radius.

■ Many computer programs can create pie charts. However, you still need to know how to create them by hand.

# AS/A2 Kite diagrams

A kite diagram is a useful technique for displaying the changing characteristics of flora and/or fauna across a study area. You are most likely to use a kite diagram when investigating vegetation succession. Succession is concerned with the way in which different species of plants dominate in a given area as the environmental conditions change. For example, you might investigate the way in which plant species change as you move inland across sand dunes.

## Drawing a kite diagram

1 Use the *x*-axis to display distance, usually across a **transect**. This data should be shown in metres. Choose the scale carefully and practise in rough if necessary.
2 Use the *y*-axis to display concentrations of each species measured in a **quadrat**. The name of the particular species should be included. This can be shown as an actual total number or more likely as a percentage. You should plot all the species found. This might be hard to fit on a graph, however, especially if there are more than 10 in your sample. Consider grouping similar types of species together if this happens. You may need help with this.
3 For each species' data, draw an appropriate scale. Practise first if necessary. Data should be plotted twice; first above zero and second below zero. This gives a symmetrical look to the graph, hence the term 'kite'.
4 Plot the data on the graph and then join up the dots. Add a key if you are displaying more than one species.
5 If no species exist at that point in the transect, use a thin black line or a blank space.
6 Label the axes and add a suitable title.
7 At the top of the graph, it is useful to draw a sketch section showing the changing physical characteristics across the transect. For example, in a sand dune transect, draw a sketch (with some element of scale) showing its changing characteristics.

### In this section you will learn:

1 what a kite diagram is used for
2 how to create a kite diagram
3 how to interpret a kite diagram
4 key points about the use of the technique.

### Key terms

**Transect:** a line along which data is collected. A road out of town or a line from cliffs to shoreline are good examples of transects
**Quadrat:** often a 1 m square, usually made of wooden poles or a metal square. It can be used for estimating vegetation content as part of a sampling strategy.

### Tip

This technique is mainly used in succession studies. A kite diagram works by showing how species concentrations change across a study area.

### Investigating species variation across sand dunes

*Case study*

As part of her studies into plant succession on sand dunes, a student collected data of the plant species across her local sand dunes.

#### Methodology

She set up a 100-m transect across a sand dune system. Using a trundle wheel, she stopped every 10 metres and dropped a quadrat where she stood. She then sampled the vegetation within the transect by estimating the percentage cover of each type of vegetation. For the percentages, she used multiples of five to estimate vegetation cover. She used a chart to identify the different species. The data she collected is shown at the top of the next page.

## How to draw a semi-log graph

1 Calculate the range of your data set (by subtracting the lowest from the highest value). This will help to determine the number of cycles needed on the logarithmic axis.

2 Remember that the cycles cannot start at zero. The first cycle has to start with some division or multiple of 10. For example, if you started with 10, your first cycle would be 10–100; or if you started with 0.01, your first cycle would be 0.01–0.1 and so on. The y-axis should be labelled appropriately.

3 The x-axis uses an arithmetic scale and should be appropriately labelled. For example, if you were showing change over time this would have an equal interval change in, e.g. years along the x-axis.

4 Carefully plot the points by matching up the x-axis variable with the y-axis variable. Care should be taken here, as it is very easy to make mistakes with the logarithmic scale. Practise with a light pencil to begin with and have your work checked.

5 Once all the points are plotted, join them together. Remember to add an appropriate title.

**Case study**

A student conducted some research into the growth of employment in computing and related industries. She was interested to see how this compared with change in employment in other sectors of the economy. Here is the estimated data she collected on employment in computing and related industries in the UK from 1970.

| Year | 1970 | 1975 | 1980 | 1985 | 1990 | 1995 | 2000 | 2005 |
|---|---|---|---|---|---|---|---|---|
| Total employed | 9,850 | 14,800 | 69,700 | 140,350 | 220,850 | 483,500 | 750,300 | 1,239,500 |

**Take it further**

Visit www.printfreegraphpaper.com to get a range of free graph paper for your studies.

**Summary**

■ Logarithmic scales are useful for showing rate of change; a steeper line indicates a faster rate of change.

■ Take care when plotting on a log scale as it is very easy to make mistakes.

■ Zero cannot be plotted on a logarithmic scale.

■ Negative and positive values cannot be displayed on the same logarithmic scale.

**Tasks**

1 Using a piece of semi-log graph paper, plot the figures for employment change in computing and related industries in the UK from 1970 to 2005.

2 Join the points with a ruler. Add a title and label both axes.

3 Describe the changes shown.

**Extension**

4 Comment on the usefulness of this technique for displaying such data.

5 Suggest an alternative technique and assess the strengths and weaknesses of the semi-log graph compared to this other technique.

## Measures of central tendency: mean, median and mode

*GCSE AS/A2*

These are probably the easiest statistical analysis techniques you will come across. This is not to underestimate their importance, though, as they are used widely in geography in many different topic areas. You should also remember that they are easily confused with each other. This could cost you marks in an examination if you get it wrong. The mean and median are also used in much more complex techniques later, so make sure you understand what these techniques are used for and how they help to analyse data.

### Mean, median and mode

You are almost certain to have come across these terms in your mathematics lessons. However, geographers also use these calculations when analysing data from **primary** or **secondary** sources. We use these techniques when we want to make a summary statement about a set of data, by giving the mid-value of the set or by stating the most frequently occurring value.

#### Mean

The mean is short for 'arithmetical mean' and is the geographer's most used measure of central tendency. It is common (but not really accurate) to substitute this term for the word 'average'.

Calculating the mean is straightforward. Do not be put off by the formula below. You simply add up all the values in your data set and divide by the number of values in that set.

---

**In this section you will learn:**

1. the basic meaning and calculation of each technique
2. the difference between the three measures of central tendency
3. key points about the use of each technique
4. the concept of the normal distribution curve.

---

**Key terms**

**Primary data:** information collected in the field by an individual or group, such as a pedestrian count in a town centre or a pebble survey on a beach. It is unique, because nobody has ever collected it before you in that time and/or in that place.

**Secondary data:** information that already exists and has been collected by another individual or group, such as weather data or census data. You come across secondary data mainly in textbooks, using the internet or in maps.

**Rank ordering:** putting the values from highest to lowest in order of size, with the largest at the top.

**Frequency:** how often something occurs. It can be expressed as a rate or as absolute figures. For example, your bus might come every 20 minutes. The frequency could be expressed as three per hour. Alternatively, if there are 12 pebbles in your sample which all have a long axis measurement of 4 cm, the frequency of that size pebble is 12.

---

**Task**

1. Using the formula provided on the right, calculate the mean rainfall in July of the 11 weather stations across north-east England, shown below.

| Station | 1 | 2 | 3 | 4 | 5 | 6 |
|---------|-----|-----|-----|-----|-----|-----|
| | 105 mm | 90 mm | 63 mm | 44 mm | 90 mm | 65 mm |

| Station | 7 | 8 | 9 | 10 | 11 |
|---------|-----|-----|-----|-----|-----|
| | 110 mm | 90 mm | 45 mm | 68 mm | 78 mm |

---

**Formula**

$$\bar{x} = \frac{\Sigma x}{n}$$

**Unravelling the formula:**
$\bar{x}$ = Arithmetic mean
$\Sigma$ = sum of
$x$ = observed values
$n$ = number in the sample

## Median

The median is used to calculate the mid-point in the data set. It is just as easy to calculate as the mean. You have to put all values in **rank order** and select the middle value. This is straightforward if there is an odd number of values. In the example above, there are 11 values. When these are rank ordered, you simply have to select the sixth value because this leaves five values on either side and it represents the median (middle value).

When there is an even number of values in the data set, and therefore no middle value, you have to perform another short calculation. Rank order the values in exactly the same way as before, but this time you total the middle two values and then divide by two (i.e. you calculate the mean of the middle two values): this gives the median value of the data set.

## Mode

Mode refers to the **frequency** with a data set and is probably the simplest measure of central tendency. To calculate this you have to note the value that occurs most frequently in the data set. In a small data set like the one above, this is very easy as there are only 11 values. It becomes much easier to make errors when you have large amounts of data to work with. You might also find that there is more than one modal value. If there are two modes, occurring in different areas of the data set, the correct term is bi-modal.

Another use of mode can be found when analysing categories. For example, in a questionnaire, you might be interested in seeing if a particular group gave a similar response. For this you could look at the modal class.

Meteorologists use a weather station to collect data on various aspects of the weather, including rainfall totals in a location over a given time period. Rainfall is usually measured in millimetres

**Tip**

Mean is used later with more complex techniques such as standard deviation and standard error of the mean *(SE)*. Make sure you are absolutely sure how to calculate the mean. If not, you will find these other analysis techniques much harder.

**Tasks**

2   Using the same example from above, calculate the median for the data set.

3   Now work out the mode.

4   Describe how the mean, median and mode differ for this data set.

# The normal distribution curve

Mathematicians noticed the normal distribution curve, also called the 'bell-shaped curve' or 'bell curve', a long time ago. It is quite a simple concept and can be used to help assess the reliability of data collected. It is important to understand the basics of the normal distribution curve, as the principles here underpin many of the statistical tests you will go on to use. Look at the following example on page 93.

**Key points**

**Mean:** the mean is particularly useful when the data has a small range. However, it is heavily influenced by extreme values.

**Median:** this is not as arithmetically sound as the mean, as it is not actually based on the figures in the data but on the rankings. However, unlike the mean, it is not affected by extreme values.

**Mode:** this technique is not useful with data that has no repeated figures as it is entirely based on frequency, i.e. how many times the item of data occurs.

## Measuring pebbles on a beach

If you took a sample of 100 pebbles on a beach and measured the long axis of each pebble (to the nearest half centimetre), you could calculate the mean of those pebbles using the formula for the mean (page 91). If you then plotted the data on a histogram (page 67), in a normal distribution you would find that the mean was also the most frequently occurring value (mode). Also, you should find that the pebbles decrease in frequency either side of the mean at the same rate. In this example the mean, median and mode are all 4 cm, which is exactly what we would expect in a normal distribution. The curve shown below can be drawn after first constructing a histogram. Once the histogram has been constructed a cross can be drawn at the top of each bar. Subsequently joining these crosses up will produce the curve.

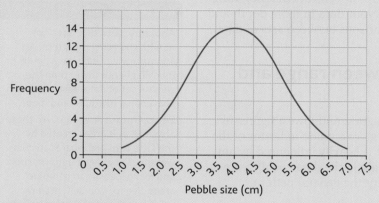

Mean, median and mode.

The larger the sample you take, the more likely it is that your data will reflect the normal distribution curve. The steeper the curve, the more clustered the data is around the mean and vice versa. If your data conforms to the normal distribution curve, you can use parametric statistical tests for further anaysis of the data (page 103).

### Key point

**What is skewness?**

Sometimes there is a clear skew in the data, when the data does not conform to the normal distribution curve. The skewness (as it is correctly termed) can be either to the left or right. In the sketches below, both types of skew are shown. If your data has either of these types of skew, you should only use nonparametric statistical tests on your data (page 108).

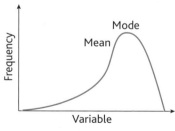

A negatively skewed distribution        A positively skewed distribution

### Tasks

5   What is the normal distribution curve?

6   How does the normal distribution curve differ from a skewed curve?

7   Write down some examples of the sorts of data that might show:
    i  a positive skew        ii  a negative skew.

### Summary

■ In the negative skew, the mode lies to the right of the mean and vice versa for the positive skew. The greater the difference between the mean and mode, the greater the skew is likely to be.

# Measuring dispersion 1: range and interquartile range

These techniques are used to analyse the dispersion, or spread of data. Read the section on mean, median and mode (pages 91–92) first because these techniques are linked. Calculating the range and interquartile range is quite straightforward.

The measures of central tendency have their own limitations, discussed earlier. Using range and interquartile range gives more information about how broadly the data is spread around the mean or median. Used together with these measures of central tendency, range and interquartile range help you analyse your data in more depth and enable you to make more detailed comparisons between sets of data.

## What is the difference between range and interquartile range?

The range is the difference between the highest and lowest figures in your data set. For example, on its own, it gives a crude indicator of the spread of data. However, like the mean, it is heavily influenced by any extreme values. In other words, an **anomaly** in the data could create a very large range and therefore misrepresent the general pattern.

The interquartile range (IQR) is simple to use, but requires a formula to work it out. Unlike the range, though, it takes away any extreme values (and therefore anomalies) in the data set. As with many other formulae used in geographical analysis, do not be put off. Carefully follow each stage from start to finish.

Look at the formula, right.

Now let's take some climate change data to show you how to work out the IQR.

### In this section you will learn:

1 the difference between range and interquartile range
2 how to interpret your findings
3 how and when to apply to apply these techniques.

### Key terms

**Anomaly:** data that does not fit with the general pattern. An anomaly will usually be a figure much higher or lower than the general set of data. Anomalies can also appear in graphs and maps.

**Super Output Areas (SOA):** these are small unit areas created by the Office for National Statistics website. They are of uniform size and make it easier to compare areas. See for yourself: http://neighbourhood.statistics.gov.uk

### Formulae

Upper quartile (UQ) = $\dfrac{n + 1}{4}$

Lower quartile (LQ) = $\dfrac{n + 1}{4} \times 3$

IQR = UQ − LQ

$n$ = number of items in the data set.

### Tip

**Working out the range:**

For example, the age range of students in an 11–18 secondary school is 7 years. The range is calculated by subtracting the lowest value from the highest.

## Case study

### Investigating climate change data

Here are some figures that show average temperature (in °C) for 2007 recorded at 15 different positions around the world. The figures compared against mean data recorded between 1951 and 1980 as the base level. Climatologists have shown this on the diagram on page 95.

3.2, 0.2, −1.4, 2.4, 3.8, 2.9, 0.8, 1.1, 0.4, 3.1, −0.8, 0.9, 2.1, −1.6, 3.9.

The figures above show no real pattern on their own. They do confirm information in the map on page 95, which seems to show quite a large difference in temperature change around the globe. Some of the temperatures show large increases, but other places show temperatures below the 1951–80 mean data. Can interquartile range help us to make sense of this?

### Key point

When using the interquartile range, it is more appropriate to use the median as the accompanying measure of central tendency for the data set. This is because the data is being placed into rank order.

Follow the sequence carefully:

1 First, put the data in rank order, starting with the highest first:
3.9, 3.8, 3.2, 3.1, 2.9, 2.4, 2.1, 1.1, 0.9, 0.8, 0.4, 0.2, −0.8, −1.4, −1.6.

2 Find the upper quartile, using the formula:
The number in the data set is 15, so $n = 15$.

3 The answer is $\dfrac{(15 + 1)}{4} = 4$. This part is very important.
**The answer is not 4.** We now need to identify the fourth figure in the rank order, which is 3.1. Go back and check this in part 1.
Use the formula for the lower quartile.
The answer is $\dfrac{(15 + 1)}{4} \times 3 = 12$. **Again, the answer is not 12.** We now need to identify the twelfth figure in the ranked data set, which is 0.2. Go back and check this.

4 The final calculation requires us to subtract LQ from UQ:
3.1 - 0.2 = 2.9
How would you interpret this figure? Look at the climate change map and see if your IQR score confirms what the map seems to show, i.e. does the map suggest quite a large variation in temperature change? What reasons could you offer for this?

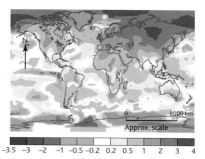

Average temperature in 2007 compared with 1951–80 mean data. This sort of data is used by climatologists to investigate climate change

## Tasks

1 Using the information provided in the table below and the formulae on the opposite page, calculate the median, range and interquartile range for city 1 and city 2.

2 What does the data analysis suggest about the differences between average weekly earnings per household in the two cities?

## Investigating differences in affluence between two British cities

*Case study*

A student compared household incomes by **Super Output Area (SOA)** in two different British cities.

He already knew that towns and cities are places with varying amounts of wealth. Some areas are rundown, while others are very wealthy (affluent). People who are more affluent tend to live together in the same parts of the city. Different cities also contain different levels of wealth. It is possible to compare differences in wealth both within cities and between cities.

The student did some research using local area statistics. He collected data on average weekly income by household in various parts of two cities.

Calculating the range will tell us the size of the gap between the richest and poorest districts in both cities, but it is heavily influenced by any extremes in the data. The interquartile range calculation allows us to take any extreme values in either data set out of the analysis, showing a range of figures without any anomalies in the data.

| SOA | 1 | 2 | 3 | 4 | 5 | 6 | 7 | 8 | 9 | 10 | 11 | 12 | 13 | 14 | 15 | 16 | 17 | 18 | 19 |
|---|---|---|---|---|---|---|---|---|---|---|---|---|---|---|---|---|---|---|---|
| City 1 Average weekly income (£) | 480 | 365 | 530 | 275 | 320 | 480 | 445 | 220 | 280 | 455 | 390 | 325 | 410 | 510 | 295 | 440 | 485 | 585 | 360 |
| City 2 Average weekly income (£) | 375 | 825 | 800 | 450 | 515 | 640 | 680 | 720 | 450 | 690 | 720 | 780 | 700 | 430 | 590 | 780 | 900 | 500 | 490 |

# Measuring dispersion 2: standard deviation

AS/A2

Like most other statistical tests, standard deviation builds upon other techniques and therefore requires you to have some prior understanding of:

- calculating the mean (page 91)
- the normal distribution curve (page 93).

You have to make sure that you understand these principles before you continue. Get some help if necessary.

## How does standard deviation measure dispersion?

You might have two sets of data that produce the same mean, but have a very different range of values within them. You could then use the interquartile range to take out the extreme values and give you a clearer idea of the spread of data. Standard deviation is one additional statistical tool that produces a figure indicating the extent to which the data is clustered around the mean.

## How standard deviation links to the normal distribution curve

You should now know that the normal distribution curve assumes that the data in your sample follows a simple distribution around the mean. The standard deviation figure gives important information in relation to this. It indicates the shape of the normal distribution curve. If the final figure is large, it suggests a wide spread of data around the mean, and a flatter, wider normal distribution curve. If the standard deviation is small, it suggests a steep and narrow normal distribution curve and a narrow spread of data around the mean. In other words, a smaller standard deviation score suggests a more reliable mean. It is also useful for comparing two sets of data that may have similar means, but quite different ranges of data within each set.

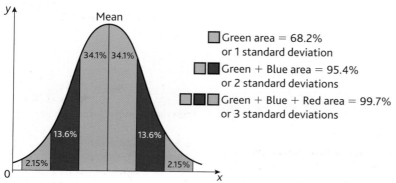

Standard deviation within a normal distribution curve.
Can you see the basic assumption of standard deviation?

### In this section you will learn:

1 how standard deviation is used to measure dispersion

2 how to apply standard deviation to a set of data

3 how standard deviation links to the normal distribution curve.

### Tip

**Sample size: how big is big enough?**

As with all statistical techniques, a general rule is that the larger the sample size, the more accurate the data produced by the sample is likely to be. However, you should also be aware that some techniques have maximum and minimum numbers of recommended sample data. Consult the individual criteria for each technique.

For more on sampling techniques, see page 27.

### Key point

**Making sense of the standard deviation calculation**

In a normal distribution:

- 68.2 per cent of the data should lie within +/- 1 standard deviation of the mean
- 95.4 per cent of the data should lie within +/- 2 standard deviations of the mean
- 99.7 per cent of the data should lie within +/-3 standard deviations of the mean.

A low standard deviation score suggests that there are few extreme values. This makes the mean a more reliable representation of the data.

## Using standard deviation to compare rainfall data

A student wanted to compare the rainfall characteristics of a region in an east African country, Kenya, with a region in a western European country, England. He noted that both had similar mean annual rainfall figures, but Kenya's annual rainfall figures seemed to show a great deal of variation over the years. He decided to investigate by comparing the standard deviation of both places' rainfall data. He selected 12 years for his comparison (top right). The figures are given in millimetres. First, look at the formula shown on the right.

Let's calculate the standard deviation for north-west England. The easiest way to complete the calculation is by creating a table:

| Year | Rainfall ($x$) | $x - \bar{x}$ | $(x - \bar{x})^2$ |
|------|------|------|------|
| 1 | 701 | 32.5 | 1,056.25 |
| 2 | 441 | −227.5 | 51,756.25 |
| 3 | 806 | 137.5 | 18,906.25 |
| 4 | 657 | −11.5 | 132.25 |
| 5 | 552 | −116.5 | 13,572.25 |
| 6 | 669 | 0.5 | 0.25 |
| 7 | 481 | −187.5 | 35,156.25 |
| 8 | 588 | −80.5 | 6,480.25 |
| 9 | 805 | 136.5 | 18,632.25 |
| 10 | 798 | 129.5 | 16,770.25 |
| 11 | 803 | 134.5 | 18,090.25 |
| 12 | 721 | 52.5 | 2,756.25 |
| | $\Sigma x = 8022$ | $\bar{x} = 668.5$ | $\Sigma(x - \bar{x})^2 = 183309$ |

1  Add up all numbers in the rainfall column (8,022).

2  Work out the mean $\dfrac{8,022}{12}$.

3  Subtract the mean from each figure to complete the $x - \bar{x}$ column.

4  Square each figure (i.e. multiply it by itself) to complete the $(x - \bar{x})^2$ column.

5  Add up all numbers in this final column. Now use this figure below.

6  Almost there! Next, you have to divide the figure by the number in the sample: $\dfrac{183,309}{12} = 15,275.75$

   Remember that the number in the sample was 12.

7  Finally, don't forget to find the square root of the figure above.

   $\sqrt{15,275.75} = 123.60$

8  Standard deviation = 123.60

### ✳Interpreting the figure

We have already calculated the mean as 668.5. This standard deviation figure tells us that in a normal distribution, 68.2% of the data lies between 544.9 and 792.1 mm. These are worked out by subtracting 123.6 from the mean to give the lower value and adding 123.6 to give the higher value.

| | NW England (mm) | Western Kenya (mm) |
|------|------|------|
| Year 1 | 701 | 389 |
| Year 2 | 441 | 786 |
| Year 3 | 806 | 990 |
| Year 4 | 657 | 1,195 |
| Year 5 | 552 | 485 |
| Year 6 | 669 | 1,293 |
| Year 7 | 481 | 531 |
| Year 8 | 588 | 372 |
| Year 9 | 805 | 421 |
| Year 10 | 798 | 983 |
| Year 11 | 803 | 384 |
| Year 12 | 721 | 693 |

### Formula

$$\sigma = \sqrt{\dfrac{\Sigma[x - \bar{x}]^2}{n}}$$

**Unravelling the formula:**

$\sigma$ = Standard deviation

$x$ = Individual value

$\bar{x}$ = Mean

$n$ = Number in the sample

$\Sigma$ = sum of

### Tasks

1  Using the example as a guide, calculate the standard deviation for western Kenya's rainfall figures.

2  What do the two standard deviation figures suggest about the rainfall in the two places?

3  Suggest some other techniques might help analyse this rainfall data.

### Key point

Using a scientific calculator to work out the answer is acceptable (and a good way of checking your result), but it is still important to have a manual grasp of how to undertake the calculation. In an examination, you might be asked to complete a small part of it.

# Measuring dispersion 3: standard error of the mean

## A2

Standard error should not be the first statistical technique that you work with. While the standard error formula is quite simple and interpreting the findings is also straightforward, you should already have a basic understanding of sampling, the mean, normal distribution and standard deviation.

## What is the standard error of the mean (*SE*)?

No sample can claim accurately to reflect exactly the **parent population** unless the whole population is sampled.

We accept that sampled data contains error. The bigger the sample taken, the less error there is likely to be in the sample. This technique allows you to assess the likely error in the mean of the sample, i.e. it allows you to suggest the reliability of the sample mean of your data set. Before continuing, look at sampling (see page 27) and at the calculation of the mean (page 91).

The key assumption of *SE* is that the data in your parent population must conform to the characteristics of the normal distribution curve (see page 93). You also need to be familiar with standard deviation technique (see page 96) before continuing.

## Calculating *SE*

Firstly you need some sampled data. As a rule for using this technique you need at least 30 pieces of data. However, you already know that the larger the sample, the more accurate the mean is likely to be and the smaller the standard error figure will be at the end of the calculation.

### In this section you will learn:

1 the theoretical basis of standard error of the mean (*SE*)

2 how *SE* links to sampling procedures

3 how to calculate *SE*

4 how to apply *SE*.

### Key terms

**Parent population:** the same as total population. When collecting primary data it is not always possible to collect data on the whole population, so we take a sample of the parent population.

**Fluvio-glacial deposits:** materials sorted by a melting glacier. In theory, the further you travel from the place where the glacier is melting, the smaller and more rounded the deposits become.

### Case study

*Analysing sampled fluvio-glacial deposits*

A student was investigating the size and shape of **fluvio-glacial deposits** in a field study in Iceland. She took a sample of 100 pieces of material and measured along the long axis (the longest part of the material). To do this, she took a ruler and accurately measured its longest axis.

After analysing the data, she calculated the mean long axis at 3.1 cm. She then calculated the standard deviation of the same sample at 0.5 cm. She then had everything she needed to calculate the likely error in the sample.

### Key point

We clearly do not have the standard deviation of the parent population. For this, we would have to have sampled every piece of fluvio-glacial material in the area. As this is not feasible, we use the standard deviation of the sample instead. This makes it even more important to get as large a sample as possible. The larger the sample, the more accurate the standard deviation is likely to be.

Sampling in cold environments (such as Iceland, shown here) is tricky and time consuming, especially when your hands are cold! On fieldwork data collection, you must pay attention to detail and remember that your group is likely to need the data you are collecting. You must make sure it is accurate

## Tasks

1  Calculate the *SE* for this student's pebble sample by substituting the information provided in the case study into the formula below. This has been started for you.

$$SE_{\bar{x}} = \frac{\sigma}{\sqrt{n}}$$

$$SE_{\bar{x}} = \frac{\sigma}{\sqrt{100}}$$

$$SE_{\bar{x}} = \frac{\sigma}{10}$$

$$SE_{\bar{x}} =$$

You now need to interpret the figure, but before you do so check your answer with your teacher.

Next, we need to revisit our confidence levels for this test. The principle is much the same as that applied to the standard deviation calculation.

2  Complete the following statements:
   • I am 68.2 per cent confident that the actual mean of the parent population will be 3.1 cm +/− (insert your figure here).
   • I am 95.4 per cent confident that the actual mean of the parent population will be 3.1 cm +/− (multiply your figure by 2 and insert here).
   • I am 99.7 per cent confident that the actual mean of the parent population will be 3.1 cm +/− (multiply your figure by 3 and insert here).

You need to know that we work to a 95 per cent level of confidence for this type of statistical test.

3  Using the 95 per cent level of confidence, what is the likely range for the mean of the parent population?

4  Explain what happens to the *SE* figure when the number in the sample increases.

## Formula

$$SE_x = \frac{\sigma}{Rn}$$

**Unravelling the formula:**
$SE_x$ = Standard error of the mean
$\sigma$  = Standard deviation of the sample
$n$   = number in the sample

Analysing correlation 1: Spearman's test

## Spearman's Rank Correlation Coefficient (*Rs*)

Spearman's test is probably the most commonly used technique for measuring correlation. This technique can be used only to examine linear relationships. You will need to draw a scatter graph first to check whether this is the case.

Before continuing, look at the section on scatter graphs (page 60). The data used to perform the Spearman's test could have first been used to construct a scatter graph. Many of the principles discussed in the scatter graph section are further developed here.

Spearman's test is a nonparametric test. No assumptions are made about the data being normally distributed.

## When should I use the Spearman's test?

If the answer is 'yes' to the following questions, a Spearman's test can be performed to analyse your data. (You may use more than 30 pairs, but the calculation becomes more difficult):

- Are you investigating a relationship between two variables?
- Do you have 10 or more pairs of data?
- Do you have fewer than 30 pairs of data?
- Are you assuming the data is not normally distributed?

Having first constructed a scatter graph using your data, you should be able to see if there is a visual relationship between your data, i.e. if there is a correlation of some sort. If the graph you create indicates no correlation, the Spearman's test is almost certain to confirm this. Similarly, if the graph indicates a positive or negative relationship, a test is worth performing to investigate this further.

The Spearman's test will indicate the strength of the relationship as a numerical value. The answer you get at the end of the calculation should lie between $-1$ and $+1$. If it does not, something has gone wrong in your calculation and you need to start again. The closer the answer to $-1$, the greater the inference of a negative relationship and vice versa. The closer the answer to 0, the greater the inference of no relationship or a random pattern. An example of how to use the Spearman's test is shown overleaf.

### In this section you will learn:

1. when to use Spearman's test
2. how to undertake the calculation
3. how to interpret the findings
4. how to apply to a set of data.

### Key terms

**Natural increase (NI):** is the difference between birth rate and death rate. A negative figure indicates that the population is falling. NI is usually given as a percentage.

**Critical value:** your result must exceed the critical value for the test at the 95 per cent level (also expressed as the 0.05 level of significance).

**Significance:** statistically, significance is only attached to our findings if we exceed the commonly accepted 95 per cent level of confidence that the result was not a statistical chance or fluke. Any less than this and we would reject the findings and accept the null hypothesis ($H_0$).

**Rejection level:** if you fail to achieve the 95 per cent level of statistical significance, you have to reject your findings. You can still discuss the result, but you cannot reject the null hypothesis ($H_0$).

**Degrees of freedom:** in a Spearman's test, the degrees of freedom are simply the number of paired measurements in the sample. For other tests, degrees of freedom are calculated in a different way.

## Investigating the relationship between natural increase and HIV prevalence rates

A student decided to undertake a secondary data research task to investigate whether any relationship exists between the **natural increase (NI)** of a country and its percentage of HIV/Aids infection in the population. She selected 14 countries from around the world. She knew from her studies that countries with high rates of HIV/Aids also tend to have high birth rates, due to a lack of contraception and many being less economically developed. However, these countries also tend to have high death rates due to the disease wiping out so many people. She expected a negative relationship between rates of HIV/Aids and rates of NI. She suggested the following null hypothesis ($H_0$):

*'There is no relationship between NI and HIV/Aids prevalence rates.'*

Like other techniques, the best way to complete the Spearman calculation is to use a table for the main part of the calculation.

You should start by making a large copy of the table on page 102. Follow the tasks in the order suggested to make the quickest progress.

### Formula

$$Rs = 1 - \frac{6\Sigma d^2}{n^3 - n}$$

**Unravelling the formula:**

$Rs$ = Spearman's Rank Correlation Coefficient

$\Sigma$ = Sum of

$d$ = Difference in rank

$n$ = Number in the sample

### Tip

It is very unlikely that you would be expected to know this or any more complex formulae. If the Spearman's test appeared in an examination, the formula will usually be given and you are likely to have to perform some aspect of the calculation as well as interpret the result. You may also be expected to know the purpose of each statistical test.

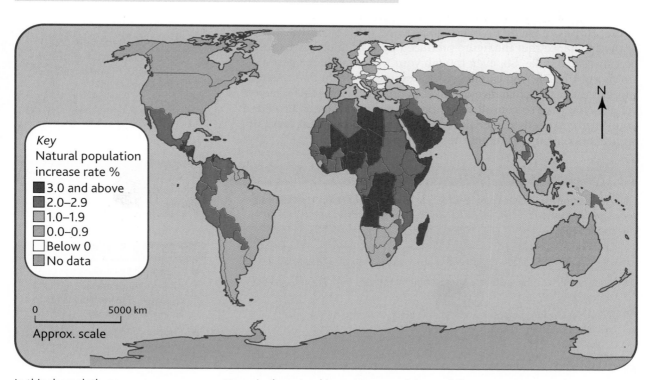

Key
Natural population increase rate %
■ 3.0 and above
■ 2.0–2.9
□ 1.0–1.9
□ 0.0–0.9
□ Below 0
■ No data

0    5000 km

Approx. scale

In this choropleth map, can you see any patterns in the natural increases around the world?

| Country | NI (%) | Rank | HIV/Aids prevalence (%) | Rank | d | d² |
|---|---|---|---|---|---|---|
| Australia | 0.10 | 9 | 0.10 | 11.5 | -2.5 | 6.25 |
| Brazil | 0.98 | 3-85 | 0.70 | 7 | -2 | 4 |
| Cameroon | 2.22 | 1 | 6.90 | 3 | -2 | 4 |
| Chad | 2.20 | 2 | 4.80 | 4 | -2 | 4 |
| Czech Republic | -0.98 | 12 | 0.05 | 13 | -1 | 1 |
| France | 0.57 | 6.5 | 0.40 | 8 | -2.5 | 6.25 |
| Germany | -0.04 | 10 | 0.10 | 11.5 | -1.5 | 2.25 |
| India | 1.58 | 3 | 0.90 | 6 | -3 | 9 |
| Japan | -0.14 | 13 | 0.01 | 14 | -1 | 1 |
| Mexico | 1.14 | 4 | 0.30 | 9 | -5 | 25 |
| Russia | -0.47 | 11 | 1.10 | 5 | 6 | 36 |
| South Africa | -5.00 | 14 | 21.50 | 2 | 12 | 144 |
| UK | 0.28 | 8 | 0.20 | 10 | -2 | 4 |
| Zimbabwe | 0.57 | 6.5 | 24.60 | 1 | 5.5 | 30.25 |

$$\Sigma d^2 =$$

## What are critical values?

You now have a result. Check this with your teacher before continuing. Once you are sure it is correct, you need to interpret the figure. However, we first need to understand **critical values**. There is always a chance that the figure produced by your analysis is a result of a statistical fluke. So, even though your figure might suggest a strong correlation, you can never be absolutely certain that this result was not the result of chance or fluke. In order to assess this issue of chance, you have to consult the critical values for the test. **Provided your figure exceeds the critical value at the 95 per cent level of significance, you can reject the null hypothesis (H₀).**

This is a one-tailed test because the student believes there will be a negative relationship (see page 109).

### Tasks

1. Rank the NI figures from highest to lowest with 1 being the highest and 14 being the lowest. This has been started for you.
2. Now rank the HIV/Aids data from highest to lowest. This has also been started for you.
3. Now subtract the second rank from the first to give the difference (d). Cameroon has been done for you.
4. Now square the differences, i.e. multiply each figure by itself to give $d^2$. Cameroon has been done for you.
5. Add up all of the values in the right-hand column to give $\Sigma d^2$
6. Now substitute into the formula:
   i $Rs = 1 - \left(\dfrac{6 \times \Sigma d^{2*}}{n^3 - n}\right)$
   *(Insert the figure produced in task 5. Check your answer with your teacher.)
   ii $Rs = 1 - \left(\dfrac{}{14^3 - 14}\right)$
   iii $Rs = 1 - ...$
   It is very easy to forget to do this next part. Your result is incorrect of you do not perform part iii!
   iv $Rs =$

### Tasks

7. Using the critical values provided, decide whether your result is statistically significant or not.

| Degrees of freedom | 0.05 level of significance | 0.01 level of significance |
|---|---|---|
| 14 13 | 0.457 | 0.646 |

8. Now go back to the student's original null hypothesis (H₀). Do you accept or reject it?
9. What conclusions do you draw from this piece of research? Use your statistical analysis to help.

### Key point

This is a one-tailed test because the student believes there will be a negative relationship (see page 109). The critical values are different for a two-tailed test, so it is important to decide at the outset which type of alternative hypothesis you will be testing.

### Key point

Whenever you square a negative, you always produce a positive.

### Summary

- Spearman's test does not imply a causal relationship, i.e. that a change in one variable leads to a change in the other.
- Choosing illogical variables may still suggest a correlation, so choose your two variables carefully.
- The test is not reliable with fewer than 10 pairs of data. With more than 30 the test becomes arithmetically difficult.

# A2 Analysing correlation 2: Pearson's test

Pearson's test – the Pearson Product Moment Correlation Coefficient (*r*) – is an alternative technique to the Spearman's test. Without a calculator or a computer program, the formula is much more complicated and it is therefore easier to make mistakes. However, it is a more accurate technique than Spearman's test because it uses actual values rather than relative ranks. Spearman's test is simpler because it uses a ranking technique (page 100). Pearson's test uses the actual data generated by your study. Pearson's test is a **parametric statistical test**.

## Case study

### Applying Pearson's test: correlating discharge with load size in a river

A student was interested in seeing if there was a correlation between bedload size (cm) and discharge (cumecs) in the River Alyn, North Wales. Through his studies, he knew that discharge tends to increase downstream in most rivers as more tributaries join the main river. He also knew that bedload size generally decreases due to erosional processes such as attrition. He expected to see a negative relationship between these two variables. He therefore proposed the following null hypothesis (H$_0$):

*'There is no relationship between discharge and bedload size.'*

*Methodology*

To calculate discharge, he first had to calculate velocity and also measure the cross-sectional area. For bedload size, he simply measured the long axis of each piece of sediment selected. At each site for data collection (in the middle of each river), he selected 10 pieces of load (clasts) and then calculated the mean long axis measurement in centimetres. Here are the results for the 12 survey sites along the river's course:

| Distance from source (m) | Discharge in cumecs (x) | Mean bedload length in cm (y) | Velocity (m/sec) |
|---|---|---|---|
| 150 | 0.20 | 4.2 | 0.15 |
| 500 | 0.31 | 4.5 | 0.20 |
| 800 | 0.26 | 3.6 | 0.45 |
| 1500 | 0.59 | 2.7 | 0.63 |
| 1950 | 0.81 | 3.1 | 0.68 |
| 2400 | 1.12 | 2.2 | 0.79 |
| 2700 | 1.01 | 2.0 | 0.63 |
| 3200 | 1.69 | 2.3 | 0.68 |
| 3700 | 1.93 | 1.7 | 0.93 |
| 4500 | 2.48 | 1.1 | 0.99 |
| 6000 | 2.21 | 0.6 | 1.01 |
| 8500 | 3.07 | 0.8 | 1.27 |

### In this section you will learn:

1 the difference between Spearman's Rank Correlation Coefficient (*rs*) and the Pearson Product Moment Correlation Coefficient (*r*)

2 how to use the Pearson Product Moment Correlation Coefficient

3 how to interpret the findings of this technique

4 other examples of where you might use this technique.

### Key term

**Parametric test:** statistical test based on the assumption that the data being analysed is normally distributed.

Sediment analysis

### Formula

$$r = \frac{\Sigma(dx \cdot dy)}{\sqrt{\Sigma d^2x \cdot \Sigma d^2y}}$$

**Unravelling the formula:**

*r* = Pearson's Product Moment Correlation Coefficient (*r*)

*d* = deviation from mean

Σ = sum of

√ = square root

· = this is an alternative way of indicating multiply.

Like other techniques, the next stage is best completed using a table.

| Discharge in cumecs (x) | Bedload size in cm (y) | Deviation of x from mean of x (dx) | Deviation of y from mean of y (dy) | $d^2x$ | $d^2y$ | dx.dy |
|---|---|---|---|---|---|---|
| 0.20 | 4.2 | 1.11 | −1.8 | 1.2321 | 3.24 | −1.998 |
| 0.31 | 4.5 | 1.00 | −2.1 | 1 | 4.41 | −2.1 |
| 0.26 | 3.6 | 1.05 | −1.2 | 1.1025 | 1.44 | −1.26 |
| 0.59 | 2.7 | 0.72 | −0.3 | 0.5184 | 0.09 | −0.216 |
| 0.81 | 3.1 | 0.5 | −0.7 | 0.25 | 0.49 | −0.35 |
| 1.12 | 2.2 | 0.19 | 0.2 | 0.0361 | 0.04 | 0.038 |
| 1.01 | 2.0 | 0.3 | 0.4 | 0.09 | 0.16 | 0.12 |
| 1.69 | 2.3 | −0.38 | 0.1 | 0.1444 | 0.01 | −0.038 |
| 1.93 | 1.7 | −0.62 | 0.7 | 0.3844 | 0.49 | −0.434 |
| 2.48 | 1.1 | −1.17 | 1.3 | 1.3689 | 1.69 | −1.521 |
| 2.21 | 0.6 | −0.9 | 1.8 | 0.81 | 3.24 | −1.62 |
| 3.07 | 0.8 | −1.76 | 1.6 | 3.0976 | 2.56 | −2.816 |

$\bar{x} = 1.31$  $\bar{y} = 2.4$

$\Sigma d^2x = 10.0344$  $\Sigma d^2y = 17.86$  $\Sigma dx \cdot dy = -12.195$

$d^2x$ is calculated by multiplying the dx figure by itself. This is the same for $d^2y$.

Next, you simply have to substitute the figures into the formula:

$$r = \frac{\Sigma(dx.dy)}{\sqrt{\Sigma d^2x \cdot \Sigma d^2y}} \qquad r = \frac{-12.195}{\sqrt{10.0344 \times 17.86}}$$

$$r = \frac{-12.195}{13.387098} \qquad r = -0.911 \text{ (2 dp)}$$

For this test, in order to calculate the degrees of freedom we use the formula n-1, where n is the number in the sample. For this data set there are 11 degrees of freedom. Remember that we can only reject the null hypothesis ($H_0$) if our result **exceeds** the critical value. Let's have a look:

| Degrees of freedom | 0.05 level of significance | 0.01 level of significance |
|---|---|---|
| 11 | 0.476 | 0.634 |

We can clearly see that our figure of −0.911 exceeds the critical value of 0.476 at the 0.05 level of significance. It also exceeds the 0.01 level of significance. From this we can reject the null hypothesis ($H_0$) at the 99 per cent level of confidence. There is clearly a negative relationship between discharge and bedload size, ie. as discharges increases, bedload size decreases in this study. Remember, though, that we cannot suggest that one variable causes a change in the other. Like the Spearman's test, this test does not imply a causal relationship.

**Tasks**

1 Using a similar layout to the style above, test the relationship between velocity and discharge using the data provided by the same study. Remember to start with a null hypothesis ($H_0$).

2 Interpret your findings using the critical values in Appendix 5.

3 Do you accept or reject the null hypothesis ($H_0$)? Give reasons for your choice.

# A2 Comparing means: the student's *t*-test

The formula for working out the student's *t*-test looks extremely complex, but by concentrating and following the rules with these sort of equations, obtaining the final figure is easier than it looks – so do not be put off. The key lies in carefully substituting your data into the formula in a logical and consistent fashion.

## What is the student's *t*-test?

The student's *t*-test is used to determine whether the data from your two samples have come from the same population or not. It is a parametric test. It therefore works on the assumption that data are normally distributed. It works by comparing the means of two data sets, in order to identify whether significant differences exist between two data sets. The alternative technique to this is the Mann Whitney *U*-test (for nonparametric data). If you are not sure which test to use with your particular data set, ask your teacher.

If the answer is 'yes' to the following questions, you can use the student's *t*-test to analyse your data:

- Are you investigating the difference between two samples of data?
- Is the data parametric (normally distributed)?
- Is the data on an **interval scale**?
- Are there fewer than 30 pieces of data in the sample?

The null hypothesis ($H_0$) should always be phrased in the following way:

There is no significant difference between the two samples.

First, look at the formula, right. Do not be put off!

### In this section you will learn:

1 the difference between the student's *t*-test and Mann Whitney *U*-test

2 examples of where you apply the test

3 how to apply the test to fieldwork data

4 how to interpret the findings.

### Tip

The student's *t*-test is also referred to as the 'unpaired *t*-test'. There is also a 'paired t-test', but this is not covered in this book.

### Key term

Interval scale: a good example of an interval scale is the Fahrenheit or Celsius scale. Here each increment in the scale is matched by an equal increment in temperature. However, you would not be able to say 20 degrees is twice as hot as 10 degrees. Also zero does not mean that nothing has been recorded. It still equates to an actual temperature.

### Case study

## Applying the student's *t*-test: Investigating microclimates

A group of students wanted to see if some characteristics of a microclimate existed in their school copse (small wooded area). The first part of their study was conducted on a bright summer's afternoon in the UK. They decided to focus on temperature variation to begin with.

### Methodology in brief

They used a random number table (see Appendix 5) to generate 10 locations within the copse and 10 locations outside the copse. They checked that their thermometers all displayed the same temperature at the start of the experiment by leaving all 20 together in the same place for 30 minutes. The 20 students stood at their agreed places inside and outside the copse and began the experiment.

### Formula

$$t = \frac{|\bar{x} - \bar{y}|}{\sqrt{\dfrac{(\Sigma x^2/n_x) - \bar{x}^2}{n_x - 1} + \dfrac{(\Sigma y^2/n_y) - \bar{y}^2}{n_y - 1}}}$$

**Unravelling the formula:**

$|\bar{x} - \bar{y}|$ = difference between 2 sample means

$t$ = student's t-test

$\bar{x}$ = mean of *x*

$\bar{y}$ = mean of *y*

$\sqrt{\ }$ = square root

$\Sigma$ = sum of

$x^2$ = square of *x* (multiply *x* by itself)

$n_x$ = number in the sample.

They each held their own thermometer at shoulder length for 10 minutes and then recorded the temperature in degrees Celsius.

The students agreed the following null hypothesis ($H_0$):

*'There is no significant difference in temperature between the mean of the sample taken in the copse and that of the sample taken outside the copse.'*

Let's begin the analysis.

Microclimate in a wooded area:

■ Temperatures tend to be lower where there is a thick canopy.

■ Humidity levels tend to be higher as plants transpire.

■ Wind speeds tend to be lower. Can you suggest reasons why?

## Tasks

1  Make a copy of the table and calculate the figures for the column $y^2$ in the same way that $x^2$ has been calculated.

| Temperature inside copse (°C) ($x$) | Temperature outside copse (°C) ($y$) | $x^2$ | $y^2$ |
|---|---|---|---|
| 19.3 | 21.7 | 372.49 | |
| 21.4 | 25.8 | 457.96 | |
| 17.9 | 20.1 | 320.41 | |
| 20.8 | 22.1 | 432.64 | |
| 19.9 | 26.4 | 396.01 | |
| 19.7 | 19.9 | 388.09 | |
| 21.9 | 22.8 | 479.61 | |
| 17.1 | 21.6 | 292.41 | |
| 18.9 | 24.3 | 357.21 | |
| 20.2 | 19.1 | 408.04 | |

2  Add up the $y^2$ column to give $\Sigma y^2$. $\Sigma x^2$ has already been done.
$\Sigma x^2 = 3{,}904.87$
$\Sigma y^2 =$

3  Calculate the key figures for ($y$).

Key figures for ($x$)
$\Sigma x = 197.1$
$\bar{x} = 19.71$
$n_x = 10$

Key figures for ($y$)
$\Sigma y =$
$\bar{y} =$
$n_y =$

## Tasks

4  Now you should have enough information to substitute your data into the formula. This has been started for you:

$$t = \frac{|\bar{x} - \bar{y}|}{\sqrt{\dfrac{(\Sigma x^2/n_x) - \bar{x}^2}{n_x - 1} + \dfrac{(\Sigma y^2/n_y) - \bar{y}^2}{n_y - 1}}}$$

$$t = \frac{|19.71 - 22.38|}{\sqrt{\dfrac{(3904.87/10) - 388.4841}{10 - 1} + \dfrac{(\Sigma y^2/n_y) - \bar{y}^2}{n_y - 1}}}$$

$$t = \frac{2.67}{\sqrt{\dfrac{2.0029}{9} + \dfrac{(\Sigma y^2/n_y) - \bar{y}^2}{n_y - 1}}}$$

$$t = \frac{2.67}{\sqrt{0.2225 + \dfrac{(\Sigma y^2/n_y) - \bar{y}^2}{n_y - 1}}}$$

5  Complete the calculation by substituting in the data for y.

6  What is your student's t-test calculation for these two data sets?

**Tip**

Look carefully at how each part of the equation for the x values has been worked out. You don't have to understand why the equation is set out as it is, but you do have to make sure that you can accurately perform the calculation.

### Interpreting the student's t-test score

You should now have a figure for your student's t-test analysis, but on its own it means nothing. We now need to interpret the figure, in order to see whether the difference between the means of the two data sets is statistically significant. Before doing this, you need to decide whether this is a one- or two-tailed test (see page 109). As the students expected the temperatures to increase outside of the copse, this is a one-tailed test.

For this test, as the two samples are equal size degrees of freedom is $n - 2$. In this case then there are 18 degrees of freedom. As with all other statistical tests, we use the commonly accepted figure of 0.05 for the minimum level of significance. If the calculated t value is less than the critical value at the 0.05 level of significance then you must accept the null hypothesis ($H_0$).

**Key point**

If the sample sizes are not equal, the degrees of freedom are calculated by $(n_x - 1) + (n_y - 1)$.

**Summary**

This test is used to see if differences exist between the means of two data sets. The data must be normally distributed and the test is limited to 30 pieces of data in each sample. There does not have to be an equal number in each sample.

Some other possible applications of Student's t-test:

- examining wealth indicators (such as car ownership rates) in two different towns
- investigating soil characteristics (such as acidity or moisture content) in two different areas of land use
- comparing vegetation characteristics (such as tree height) in two forests.

## Tasks

7  Using the extract from the critical values table, decide whether to accept or reject the null hypothesis ($H_0$). Give reasons for your answer.

| Degrees of freedom | 0.05 level of significance | 0.01 level of significance |
|---|---|---|
| 18 | 1.73 | 2.88 |

8  What do your findings suggest about the temperature variation in this microclimate study?

Next, she substituted data into the formula:

$$U_x = N_x . N_y + \frac{N_x(N_x + 1)}{2} - \Sigma r_x$$

$$U_x = 10 \times 10 + \frac{10 \times (10 + 1)}{2} - 120$$

$$U_x = 100 + \frac{110}{2} - 120$$

$$U_x = 100 + 55 - 120$$

$$U_x = 35$$

## Tasks

2   You now have the figure for $U_x$. Use the same method of calculation to work out the figure for $U_y$. Here is the formula:

$$U_y = N_x . N_y + \frac{N_y(N_y + 1)}{2} - \Sigma r_y$$

3   You now need to select the smaller of the two figures and use the critical values table (see Appendix 5) to decide on the statistical significance of your result. For this test, if your result is **equal to or smaller** than the critical value at the 0.05 level of significance, then you can reject the null hypothesis ($H_0$).

For 10 figures in each sample, the critical value at the 0.05 level of significance is 23.

4   Do you accept or reject the null hypothesis ($H_0$)? Give reasons for your answer.

5   What do these findings suggest about the difference in traffic flow before and after the retail development in this study?

What effect does new development have on traffic in a town centre?

**Summary**

Mann Whitney $U$-test can be used to compare any two data sets that are not normally distributed. As long as the data is capable of being ranked, then the test can be applied.

Other possible uses:

- Investigating differences in questionnaire responses relating to a new development.
- Investigating differences in species diversity near to footpaths.
- Investigating differences in vegetation cover between two different slopes.

# A2  Examining difference: chi-squared ($\chi^2$)

As with other skills, if you focus hard from the beginning and carefully follow the stages, the chi-squared ($\chi^2$) calculation becomes much easier.

## When to use chi-squared

Chi-squared is used to examine differences between what you actually find in your study and what you expected to find. Look at the list of questions below. If the answer is yes to each question, a chi-squared test is appropriate:

- Are you trying to see if there is a difference between what you have found and what would be found in a random pattern?
- Is the data gathered organised into a set of categories?
- In each category, is the data displayed as frequencies (not percentages)?
- Does the total amount of data collected (**observed data**) add up to more than 20?
- Does the **expected data** for each category exceed four?

Rather than trying to explain all the terms above, it is better to try an example. By doing this, the test should become clear as you work through it. The example below is quite a simple one.

## Performing the calculation

You would use chi-squared if you were investigating difference between what you actually observed (by collecting primary or secondary data) and what you might normally expect to find. Here is an example from a piece of human geography research into urban development relating to the 2012 Olympics site. First, look at the chi-squared formula, right:

An artist's impression of the site of the 2012 Olympics

---

### In this section you will learn:

1  when to use chi-squared
2  how to perform the calculation
3  how to interpret the results
4  the strengths and weaknesses of this technique.

---

### Key terms

**Observed data:** the data you collected.

**Expected data:** the data you expect to find in a random pattern.

---

### Key point

Like all statistical tests, chi-squared is not used to prove a hypothesis. It is only used to disprove a null hypothesis ($H_0$).

---

### Formula

The formula
$$\chi^2 = \Sigma \frac{(O - E)^2}{E}$$

**Unravelling the formula**

$O$ = the observed frequencies
$E$ = the expected frequencies
$\Sigma$ = the 'sum of'

# Using chi-squared to analyse questionnaire responses

A student collected data on local people's viewpoints about the building of the 2012 Olympic Games venue in Stratford, east London. She was interested in seeing if viewpoints changed according to the perspectives of various different groups.

## Method in brief

She decided to collect 20 responses from each category of local person. After this, she discontinued the data collection. She knew nothing about the local population's demographic characteristics and did not try to reflect this in her study. Her questionnaire was a survey of viewpoints about the usefulness of the new Olympic developments for different groups (categories) of locals. This is the statement she posed:

'The 2012 Olympic Games development will be of benefit to the whole community of Stratford, east London.'

| 1 | 2 | 3 | 4 |
|---|---|---|---|
| Strongly agree | Agree | Disagree | Strongly disagree |

She wanted to find out if the type of local person influenced feelings about how useful the developments would be. She was particularly interested in those who felt negatively abo ut the Olympic developments (disagreed or strongly disagreed).

Here are the results of the survey. The residents had to choose which category best suited their own characteristics. Some of those questioned could have fitted into more than one category. For example, a business owner may also be a local resident. In this case the person questioned was disregarded from the study. Remember that 20 people responded from each category and she only recorded the **frequency** of negative response, i.e. those who either disagreed or strongly disagreed with the statement.

| Category (type) | Frequency of negative responses (Observed values: O) |
|---|---|
| Business owner | 4 |
| School student | 6 |
| Adult male resident | 14 |
| Adult female resident | 10 |
| Senior citizen | 16 |

The student added up the number of times a negative response to the question above was given. You can see that the categories seem to show large differences of opinion between the groups.

A glance at the results suggests that different groups have responded very differently to the questionnaire. The chi-squared calculation helps us decide if there is a statistically significant difference between the groups. You can then use the critical values to assess the likelihood of the results being a chance or fluke set of figures.

The London 2012 logo

## Key terms

**Frequency:** the number of times the piece of data appeared in a category.

**Null hypothesis ($H_0$):** this suggests that there is no significant difference between your observed and expected data.

The first task is to generate a **null hypothesis ($H_0$)**:

*'There is no significant difference between the category of local person and the frequency of negative response.'*

Now let's test this null hypothesis ($H_0$). It is much easier to start the calculation with a table.

|  | Business owner | School student | Adult male resident | Adult female resident | Senior citizen | Total |
|---|---|---|---|---|---|---|
| O | 4 | 6 | 14 | 10 | 16 | 50 |
| E | 10 | 10 | 10 | 10 | 10 | 50 |
| O − E | −6 |  |  |  |  | ---- |
| $(O - E)^2$ | 36 |  |  |  |  | ---- |
| $\dfrac{(O - E)^2}{E}$ | 3.6 |  |  |  |  | ---- |
| $\chi^2$ | 3.6 |  |  |  |  |  |

In this example, the expected data (E) is simply taken as being the mean negative frequency of response. It is calculated by adding up all of the observed data (O) and then dividing by the number of categories, i.e. 5. This gives an expected frequency of 10 for each category.

Don't forget to add up the bottom row to get the final $\chi^2$ figure.

*Interpreting the results*

You now need to make sense of this figure. You should first double check your answer with a partner or teacher.

For the next part of the calculation you need to understand degrees of freedom and significance levels. This is covered in several of the other statistical skills. If necessary, have a quick read over these now before continuing (from pages 100).

For this type of study, the degrees of freedom is calculated as n − 1, where n is the number of categories in the sample. As there are five categories, there are four degrees of freedom.

You now need to consult your critical values table (see Appendix 5) to see if you can reject your null hypothesis ($H_0$) at the 0.05 level of significance. Remember that statistically speaking, this means that you can be 95 per cent confident that these results are not a statistical fluke. Or, put more simply, there is a 5 in 100 chance that these results are a fluke, suggesting a misleading difference.

In order to reject the null hypothesis ($H_0$), your Chi-squared score must be greater than the critical value at the 0.05 level of significance.

| Degrees of freedom | 0.05 level of significance | 0.01 level of significance |
|---|---|---|
| 4 | 9.49 | 13.28 |

**Tasks**

1 Make a copy of the table, left, and add in the missing values. The figures for the 'business owner' category have been completed for you.

2 What is your answer for $\chi^2$?

3 Using the critical values provided (see Appendix 5), decide whether to accept or reject the null hypothesis ($H_0$). Give reasons for your answer.

4 What does this suggest about the views of the people of Stratford in relation to the 2012 Olympic Games?

5 What other research could be carried out to investigate this further?

**Tip**

You might use chi-squared for:
- Investigating the difference in orientation of corries in a glacial landscape.
- Investigating differences in the location of industries.
- Investigating house price variation across a town.
- Investigating how underlying rock type influences type of farming.

**Summary**

- Like all statistical techniques, chi-squared gives objective statistical significance to the results. So in this example, you are not just stating that there seems to be a difference between what different groups feel about the 2012 Olympic Games. This test allows you to state with much greater authority and certainty whether the difference is statistically significant or not.

- The test becomes a little more complicated when the observed data is not evenly spread, for example if more males were questioned than females in a questionnaire. The expected data has to reflect this.

# Interpreting photographs

You have now had some experience of using digital images in presenting your work, but you are also likely to face digital images in your examinations. There are some simple rules to follow when interpreting digital images under examination conditions:

1  The image you look at in the examination will have been carefully chosen by the examiner. There will be very specific reasons for the use of that image. It will in some way show something very specific to the part of the specification being examined. This may not be immediately obvious and may be quite subtle.

2  Identifying the key reason for the photograph requires you to ask a really basic question. You should try to get into the mind of the examiner and ask yourself: 'What does the examiner want me to see in this image?' You may also need to think about the area of your specification being examined in answering this question.

3  You may be asked to label or annotate the images. Always make sure that you carefully point directly to your target in the image. Easy marks are often lost when candidates forget to add arrows linking their annotation to the feature in the image.

4  You may be asked to identify evidence in an image. To do this you need to pay close attention to detail and accurately identify your key features.

Here is an example.

> **Tip**
>
> In marking this question, the examiner would have indicated 4 max. on the work. This is because the candidate has made more than four points worthy of credit. Though there are only four annotations, the candidate has clearly identified reasons and also explained some points.

**Case study**

In an examination with a question on town centres, students were asked to annotate the photograph identifying and explaining evidence that this town centre was in decline. The question was worth 4 marks. Here is what one student wrote. He scored full marks.

There is a lack of pedestrians which suggests that there is little demand for the services.

There do not appear to be any big brand name stores or multiple chain stores.

Most main streets in towns are pedestrianised. The fact that this street is not suggests there is a lack of demand for the services.

These two shops appear to sell low-order goods (e.g. alcohol). This suggests land rents are low because of a lack of demand for property.

This image was taken at 2 pm on a Saturday in July in Runcorn's main shopping street

Glacial trough

**Tasks**

1 Describe the key features of the glacial trough shown left.

2 What landform would you expect to see immediately behind the photographer?

**Extension**

3 With the help of evidence shown in the photograph and your own knowledge, explain how glacial troughs form.

# Satellite images

Satellite images have never been more widely used in teaching and learning. Every time your teacher uses Google Earth to show you a feature of the world, you are using satellite technology. However, satellite images have been around for many years and are most commonly used in weather studies. In order to fully understand how to use satellite images of weather patterns, you really need to have studied the basics of synoptic charts. Ideally you would have already had some lessons on weather and climate before trying to describe these images. This is explored on page 57. Two widely used images are those showing the track of a hurricane and those showing the passage of a depression.

A clearly evident anticlockwise rotation into the centre of the hurricane.

A well-developed eye.

The hurricane is several hundred kilometers in diameter.

Cloud becomes increasingly thick towards the centre of the hurricane.

Satellite image of Hurricane Katrina, taken 29 August, 2005

0        300 km

Approx. scale

In an examination, students were asked to annotate this satellite image of Hurricane Katrina to show its characteristics. One student's annotations are shown on the image above. The question was worth four marks and she scored full marks. Can you see why?

**Case study**

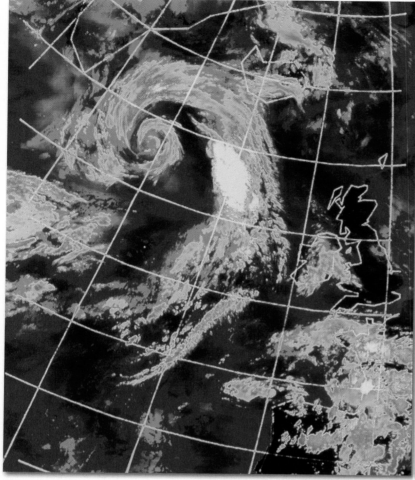

Satellite image of a depression

0       500 km

Approx. scale

## Oblique aerial photographs

In the past, these images would have been taken by aeroplanes and helicopters, but in the example below, satellite technology has been used to create this image. Unfortunately, with this sort of image it is impossible to zoom out any further. This is a limiting factor because we cannot get access to the rest of the central area on one image. Can you think of any other drawbacks?

Oblique aerial image of part of the centre of Liverpool

### Tasks

1 Describe the characteristics of the depression shown in the satellite image, left.

**Extension**

2   i   Suggest an alternative technique for presenting this data.

    ii   Comment on the advantages of this technique over a satellite image such as that shown on page 121.

3 Comment on the strengths and weaknesses of using satellite images to display information in this way.

4 Using information from the photograph below, identify evidence that suggests that this is the central area of Liverpool.

### Key point

**What is remotely sensed data?**

Remote sensing involves the use of technology (such as satellites) to gather, interpret and analyse data from places that humans would otherwise find too difficult, expensive or dangerous to obtain it from. For example, in studying global warming, if you were to use carbon dioxide ($CO_2$) data collected by a high altitude balloon, you would be using remotely sensed data. Analysing data from a weather station is also a form of remote sensing.

### Summary

- When taking photographs, practise by taking more than one picture and avoid zooming in too much.
- Always pay close attention to detail when interpreting images.
- When annotating, clearly point to your chosen features and add explanation if required.

# Using geographical information systems (GIS)

*'A geographical information system (GIS) integrates hardware, software, and data for capturing, managing, analysing and displaying all forms of geographically referenced information.'*

(*www.gis.com*)

GIS is one of the new technologies available to help you gain access to information, present data, explore patterns and even make predictions about future events.

GIS requires four different elements to operate:

1 Hardware such as computers, satellites, digital cameras and global positioning systems (GPS) are needed to capture and analyse data.

2 Software is needed to organise, categorise, sort and manipulate data. There are many computer programmes available for this. We will be referring to AEGIS 3, which is a GIS software system designed for school-based use.

3 Digital maps are usually needed for the display of data. These can come in many forms.

4 Data also comes in many forms. This can either be primary or secondary data (see page 13), depending upon what you are trying to achieve.

There is an important distinction to make here. Using the census data to generate a choropleth map of the population density of different cities within the UK does show that you can work with GIS. However, you are only using secondary data here and you are not engaged in creating your own GIS from data you have collected.

## What is AEGIS 3?

AEGIS 3 is a piece of educational GIS software that has been specifically developed to allow secondary pupils to work with GIS in a more 'hands-on' way. Using AEGIS 3, you can work with your own maps of your local area and add various types of geographical data to display specific information about your community. There are a wide range of possible topics you could investigate such as:

■ microclimate studies

■ land-use mapping

■ traffic flows

■ environmental assessments (including the addition of digital images to your map)

■ sphere of influence studies.

---

### In this section you will learn:

1 the meaning of GIS

2 the different types of GIS

3 how to get started with AEGIS 3

4 key points about the use of GIS.

### Key points

GIS comes in many forms. Here are just a few examples:

1 **Google Earth** uses satellite data to provide incredibly detailed aerial images of different parts of the world.

2 **The Meteorological Office** provides up-to-the-minute data on weather patterns in the UK.

3 **The census** provides vast amounts of information on the population of the UK. This can be manipulated to produce detailed maps and graphs.

4 **The Environment Agency** maintains up-to-date information about a range of environmental issues affecting all parts of the UK.

---

  **The Advisory Unit: Computers in Education**

The Advisory Unit produces AEGIS 3, widely used in many secondary schools

# The basics of using GIS software

The Advisory Unit provides a detailed manual and guidance on how to use the software. Further information on AEGIS 3 products and services can be accessed via **www.advisory-unit.org.uk**. In order to get started all you need is a digital map or aerial photograph and your software. The main provider of UK maps is the Ordnance Survey (OS). They produce OS MasterMaps. These **vector data** maps arrive electronically and are ready for use with AEGIS 3. The Ordnance Survey also provides free maps of your locality, covering a maximum area of 16 km². These are available at **www.ordnancesurvey.co.uk/oswebsite/getamap**. These maps are limited, though, if your study area is large. All OS digital maps are suitable for use with AEGIS 3.

Goad plans are another alternative (see page 54). These provide detailed information on services and functions within town centres. They are particularly useful for land use surveys within town. The Advisory Unit has an additional module for using Goad plans. These too are vector maps and are immediately ready for the addition of your data from your fieldwork or secondary data collection.

A final alternative is aerial images such as those produced by satellite technology. These can also be incorporated into AEGIS 3. However, these images are in **raster data** format and need to be overlaid with vector data before you can use them with AEGIS 3.

Once you have your map and your data, you can begin to produce some really interesting electronic maps using this GIS software. Here are some examples.

## Key terms

**Vector data:** data made up of lines, areas and points. This type of data is well suited to manipulation by GIS software programmes such as AEGIS 3.

**Raster data:** includes digital images or scanned-in maps. This data has to be overlaid with vector data (line, areas or points) before it can be manipulated by GIS software.

**Distance decay:** a theory widely used in migration studies. The basic idea is that people become increasingly less likely to migrate to a given place as the distance between the origin and destination increases. In the study below, the students are trying to find out if the number of visitors to Hitchin decreases as distance from the settlement increases.

## Case study

### Investigating the sphere of influence of Hitchin

As part of their investigation into sphere of influence, a group of GCSE students designed a questionnaire in school with their teacher. They were interested in seeing whether their town was more likely to be visited by people from nearby settlements. They were also interested in seeing if there was any evidence of **distance decay**. They suggested two null hypotheses (see page 112):

*'There is no relationship between the number of visitors to Hitchin and the population size of the origin settlement.'*

*'There is no relationship between distance of settlement to Hitchin and number of visitors.'*

They then used AEGIS 3 to produce a map with a combination of desire lines and proportional circles (see pages 43 and 47). The desire lines were used to show the number of visitors to Hitchin from each settlement surrounding the town. The proportional circles were used to show the population of each settlement surrounding Hitchin.

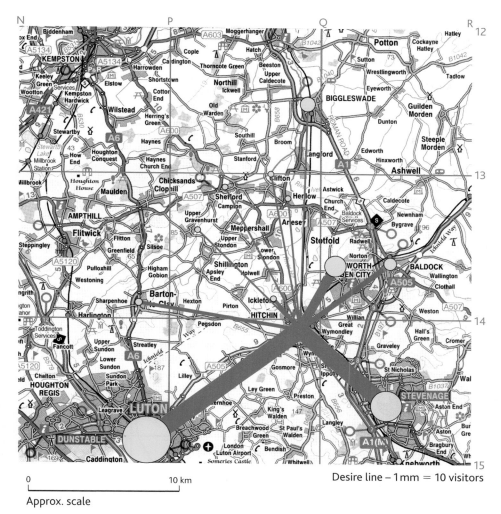

0             10 km

Approx. scale

Desire line – 1mm = 10 visitors

The sphere of influence of Hitchin, North Hertfordshire.

## Tasks

1 Using the map and data provided, describe the pattern shown by the desire lines and proportional circles.

2 Giving reasons for your answer, explain whether distance from Hitchin or size of surrounding settlement is the biggest factor affecting the number of visitors to Hitchin in this study.

### Extension

3 Suggest an alternative technique for displaying the population data for the surrounding settlements.

4 Outline the advantages and drawbacks of using GIS technology over hand-drawn techniques.

### Population of surrounding towns

|  | Settlement | Population |
|---|---|---|
| 1 | Biggleswade | 12,961 |
| 2 | Shefford | 3,336 |
| 3 | Gravenhurst | 515 |
| 4 | Barton-le-Clay | 3,488 |
| 5 | Luton | 183,000 |
| 6 | Clifton | 2,654 |
| 7 | Henlow | 3,337 |
| 8 | Letchworth | 31,418 |
| 9 | Baldock | 9,232 |
| 10 | Stevenage | 79,000 |
| 11 | Knebworth | 4,372 |
| 12 | Ickleford | 1,877 |

Population of towns surrounding Hitchin

# Investigating patterns in population structure in Carlisle

A group of A-level students were investigating the extent to which the provision of services varied around Carlisle. They started with this research question:

*'Does the age structure of the people in different parts of Carlisle affect the type of services provided in different districts?'*

To help answer this research question, they first conducted a secondary data search to find out the age structure in different parts of the city. From this, they were able to work out the mean age of the population in each ward (see page 91) and also the relative proportions of elderly people in each ward. Using their AEGIS 3 software, they chose choropleth mapping and proportional divided circles (see pages 46 and 49) as the techniques for displaying their data.

## Tasks

5   Describe the pattern of the mean age of population in Carlisle in 2001.

6   Comment on the link between mean age and the relative proportions of elderly people in Carlisle in 2001.

## Summary

■ GIS is a system that allows you to store, sort, categorise, manipulate and display geographical data through electronic media.

■ Using a GIS programme is a skill that takes time to develop.

■ When deciding how best to display your data, weigh the advantages and drawbacks of GIS over more traditional methods.

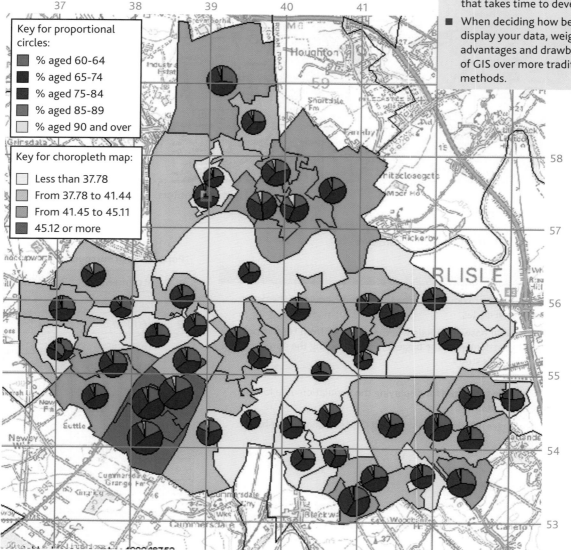

Key for proportional circles:
■ % aged 60-64
■ % aged 65-74
■ % aged 75-84
■ % aged 85-89
□ % aged 90 and over

Key for choropleth map:
□ Less than 37.78
□ From 37.78 to 41.44
▨ From 41.45 to 45.11
■ 45.12 or more

Comparing mean age with age structure for the over-60s in Carlisle in 2001

0    1 km
Approx. scale

# Using internet databases

You are more likely to have come across data~~~~~~~~~~~~~~
lessons. However, geograph~~~~~~~~~~~~~~~~~~~~ hing
secondar~~~~~

A databas~~~
informatic~~~
informatic~~~~~~~~~~~~~~~~~~~~~~~~~~~~~~ ords it
has stored~~~

Taking the~~~~~~~~~~~~~~~~~~~~~~~~~~~~~~ted
in the UK a~~~~~~~~~~~~~~~~~~~~~~~~~~~~~ the
Police Natic~~~
access to pa~~~~~~~~~~~~~~~~~~~~~~~~~~~~ntify
quickly the r~~~~~~~~~~~~~~~~~~~~~~~~~~~ y
individual in~~~~~~~~~~~~~~~~~~~~~~~~~~~~ed
to obtain stat~~~~~~~~~~~~~~~~~~~~~~~~~~ well
as investigati~~~
locations (see~~~~~~~~~~~~~~~~~~~~~~~~~~ g
the informatic~~~~~~~~~~~~~~~~~~~~~~~~~~ing
business. Thou~~~~~~~~~~~~~~~~~~~~~~~~~~ this
sort of informa~~~

While the gene~~~~~~~~~~~~~~~~~~~~~~~~~~ al
Computer, you~~~~~~~~~~~~~~~~~~~~~~~~~~~ our
local area. How~~~~~~~~~~~~~~~~~~~~~~~~~ ise
databases such~~~
**census/index.h**~~~~~~~~~~~~~~~~~~~~~~ sic
information abo~~~
**www.upmystree**~~~
geography before~~~
trying the census~~~

**In this section you will learn:**

1 what a database is

2 some examples of databases used in geography

3 how to access databases and select appropriate information

4 key points about the use of internet databases.

## Searching a database

A database such as the online census contains a vast amount of information and has many benefits. However, there are also limitations that mainly affect inexperienced users. Here are some tips on using very large databases such as the online census or the Met Office.

1 Give yourself plenty of time before starting a search such as this, especially if it is your first time. You have to be prepared to spend at least an hour searching a large database like the online census. If you are in a rush, you are much more likely to make mistakes and miss important pieces of information.

2 Be clear about what you are looking for before you start. If you are too vague at the outset, you will produce vague results. If you are too specific about what you are looking for, you may not find the record you are looking for. Get advice if you are not sure.

3 In simple terms, the less experienced you are using databases the longer you will need to practise navigating around the database.

4 All publicly accessible databases on the internet have a search facility on the home page. This allows you to search the whole database with a key term you may have. However, avoid using the search facility with general words.

For example, if you typed 'population' into the search facility of the online census, you would get an extremely large number of '**hits**', whereas if you typed in 'population in Manchester 1981', the search would produce much more specific results.

5 When you have found what you are looking for, always keep a record of where you found the piece of information. If you submit any work for marking that is not your own, you could get into serious trouble for **plagiarism**. Therefore you should source your data with an appropriate reference.

6 Once you have your data, decide how to present it. The data should be simply and clearly presented, preferably using word processing or spreadsheets. You may also be asked to present and analyse it in other forms, such as drawing graphs and performing statistical tests on the data. You should consult the index of this book in order to get help with the relevant technique.

## Key terms

**Hit:** occurs when your search term matches a record in the database. Too few hits for a general search suggests your search term was too narrow. Too many hits and it will take too long to search through them all, suggesting your search term was too broad.

**Plagiarism:** occurs when you hand somebody else's work in and claim that it is your own work. Downloading data from the internet (without acknowledging it), copying in an examination or copying somebody else's coursework are all examples of plagiarism. You can be thrown off your course for this and in certain circumstances you may be breaking the law.

Always make it clear when you have borrowed somebody else's material or ideas. You can do this by adding a reference list at the end of your work. Your teacher will help with this.

## Case study

### Producing a local area report

A group of students from Winsford in Cheshire were asked to use the Office for National Statistics (ONS) online census to produce a detailed report about the characteristics of the local area around their school. At this stage, no attempt was made to find reasons for any patterns or differences from national averages that they found. They planned to use this survey as part of an introduction to their controlled assessment (see page 9). All that they were given was the website for the ONS: **www.statistics.gov.uk**. From here they navigated to the neighbourhood statistics. On that page they were asked to input a postcode. After doing this, a large summary was produced from the database. They then had to decide how best to display this information. They decided to start with a map of the area (see page 129), downloaded from the online census website.

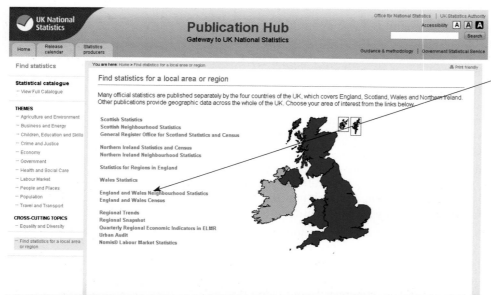

Here is the link to neighbourhood statistics on the ONS website. Once you click on this link, you will be prompted to give a postcode. Once you input this, your local area summary will be automatically produced.

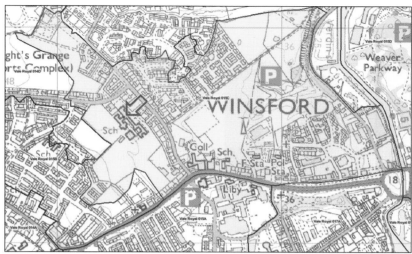

**Tip**

The UK population characteristics are now likely to be significantly different from the picture suggested by the 2001 Census. This is because the census is several years out of date. Always bear this in mind when you are writing about population change. The ONS produce some more up-to-date information on their website, though it is by no means as comprehensive as the census.

This simple map was produced by the search. The postcode is identified by the red arrow. While this gives a good, clear image of the local area, some important pieces of information are missing. Can you tell what they are?

**Tasks**

1  Visit the website **www.statistics.gov.uk** and find the neighbourhood statistics pages.
2  Write a short description of the meaning of each of the indicators shown on the right.
3  Summarise the neighbourhood statistics for this area in Winsford, Cheshire, shown on the right.

*Case study*

Here is the beginning of the students' written summary of the local area based on the online census information:

'Winsford lies in the borough of Vale Royal. It experiences higher than average levels of deprivation based on many of the available indicators. The population is mainly white (97.5 per cent based on 2006 data) and Christian (82.1 per cent). The age structure of the population is well balanced, with most being within the working population age group (16–64).'

**Key terms**

**Dependency ratio:** is the proportion of the combined numbers of under-16 and 65 year old and over (males and females) against the total population. A high dependency ratio is usually bad for the local economy, because this means it has a high proportion of children and pensioners to support who usually do not work. It is calculated in the following way:

$$\frac{\text{No. 15 and under} + \text{No. 65 and over}}{\text{Number of working age}} \times 100 = \text{dependency ratio}$$

In 1999, the UK figure was 53.55, so, for every 100 people of working age there are 53.55 people dependent upon them.

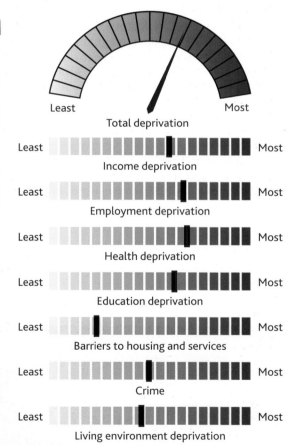

This diagram gives a very simple visual summary of the situation for the population in the area around the students' local school in Winsford. The data for Winsford is compared against least and most deprived areas nationally. What does the 'Barriers to housing and services' statistic suggest about the availability of housing around Winsford?

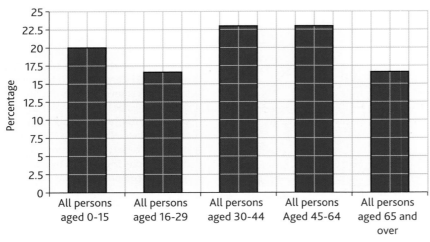

Population structure in this part of Winsford in 2006. The total population in this area in 2006 was 1,685. Can you estimate the **dependency ratio** from this bar graph and the population figure?

The health statistics for the local area were similar to the national averages on most indicators. There were a few differences worth noting, though. Teenage pregnancy rates (measured per thousand 15–17-year-old girls) were lower than national averages (37.6 in Vale Royal compared with 41.3 nationally). Also, long-term health problems were significantly higher than national averages, with 20.6 per cent of the population having some sort of long-term health problem compared with 17.9 per cent nationally.

**Tip**

Some specifications require you to experience the skill of presenting graphs, maps, text and images using ICT. By producing a report such as this using a program like Microsoft Word, you will have achieved this skill.

## Tasks

4  At the Office for National Statistics website (**www.statistics.gov.uk**), type the postcode CW7 2BT into the neighbourhood statistics search box and finish the short report about the local conditions for the people of this area of Winsford. This has been started above.

**Extension**

5  Type in your own postcode in the neighbourhood statistics part of the online census website and compare the situation in Winsford with your own local community.

6  Suggest reasons for the differences you have identified.

'Much of Winsford grew in the 1970's as inner-city families from Manchester and Liverpool were re-housed in new social housing developments. This initiative brought with it some problems though.' What might these have been?

## Summary

- Performing internet searches of any type takes time. If you rush, you are more likely to make mistakes or miss important information.

- When searching a database for records, choose your search terms carefully.

- Once you have your information, always keep a reference and if you are submitting a report, never claim other people's work as being your own.

**Researching using the internet**

It is almost certain that at some point during your GCSE or A-level course, you will be asked to do some independent research of your own. This might be part of an external examination such as the controlled assessment at GCSE. Alternatively, it might form part of the preparation for a timed examination where you are asked to explain how you used research in your course. This section has two main functions: firstly you will be shown how to undertake research using the internet; secondly you will gain some experience in answering questions about the research you have undertaken.

If your teacher has recommended some pages in textbooks that you should look at as part of your research, try to organise this before doing anything else. Nowadays, course-specific textbooks are highly focused upon your own specification and are likely to contain lots of relevant information about your topic area.

## Using internet search engines

In its simplest form, your research task might be to find out some additional information on a case study you have been looking at in class. Your teacher is likely to have already given advice and explained the processes involved in undertaking personal research. Let's now assume you are about to start a small research project of your own. Here are some tips to get you started with research using the internet:

1 Just like researching using a database, you must set aside some time for yourself. Internet research takes a lot of time to complete. In the past, you might have used a library to find information and it takes time to find a book or journal that is suitable for your requirements. Now the problem is very different. You will almost certainly get information overload, i.e. too many potential websites to look at. Choosing appropriate search terms is therefore crucial.

2 Search engines such as Google make a profit by advertising websites of companies and organisations. For you, this is a little complicated. It usually means that the first few 'hits' at the top of your search list are likely to be companies selling products. You have to ignore these. The search engine usually displays these as 'sponsored links'. You want the information and education websites, not the sales websites. This will only add time to your research, so ignore these sites.

3 When you click on the **hyperlink** to the website, have a quick browse to see if it contains suitable information. If it does contain information you are interested in, either save it as a favourite, or write down the web address. Don't spend too long reading at this stage because you will need to visit quite a few websites before you narrow down your search.

### In this section you will learn:

1 how to structure your research

2 how to use search engines

3 some of the pitfalls of internet research

4 the importance of referencing the work of others

5 key points about this technique.

### Key point

See page 128 to make sure you understand plagiarism.

### Key term

**Hyperlinks:** usually appear in blue text and are underlined on a web page. When you click on the hyperlink, it should take you directly to the page you are interested in.

4 Depending on your project, you should set aside two or three hours for initial research. Once you have visited several websites, you should now have a list of suitable ones for your research. It is now time to revisit the task that has been set and start to plan how you are going to address the requirements.

5 Remember that most search engines give you the option of searching for images, videos, maps and so on. Decide what kind of content you want to search for and instruct the search engine accordingly.

## Case study

### Generating a case study using internet research

A GCSE student was given the following task:

'Create your own case study of an active volcano. In your case study you should:

■ identify the location of your chosen volcano

■ describe the recorded history of eruptions

■ describe its characteristics and how its shape fits with the expected theory

■ describe its current status: active, dormant or extinct

■ explain how local people manage the risks associated with living around the volcano.'

Faced with this research task, she began a preliminary search using Google as her search engine. From her studies, she knew that Mount Pinatubo was a good case study because it is an active volcano and has erupted several times recently.

By typing 'Mount Pinatubo map' in the Google search box, she had a choice of maps at different scales to choose from. She selected one that showed the volcano in its regional setting.

This map shows the location of Mount Pinatubo in the Philippines. How could the student have improved her location map, particularly for somebody with no knowledge of where Pinatubo is in the world?

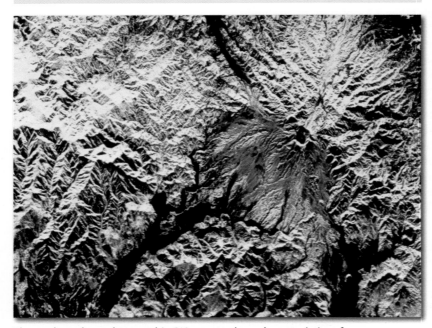

The student planned to use this GIS map to show characteristics of Mount Pinatubo. How could she have improved the layout of this image? How would you describe the characteristics of Mount Pinatubo?

The student continued her research by seeking out information about recent eruptions. In the search engine she typed in 'Mount Pinatubo'. As this brought in over 96,000 possible websites, she narrowed the search by typing in 'Mount Pinatubo eruptions'. This still revealed over 55,000 sites, so she proceeded by visiting the most popular websites. Once she started visiting these sites, other links were suggested. She kept a detailed list of each site she visited and began to build a detailed research portfolio.

## Task

1 Complete this short report on the volcano, Mount Pinatubo. In your report you should:
   - describe the recorded history of eruptions
   - describe its characteristics and how its shape fits with the expected theory
   - describe its current status: active, dormant or extinct
   - explain how local people manage the risks associated with living around the volcano.
   Your report should be no longer than 750 words.

What are the people doing in the area around Mount Pinatubo and why?

# Open-ended research

At A-level, or as part of your GCSE controlled assessment, you may well be given a much more open-ended research task, which only gives you a question and no other guidelines. It is still worth doing some background research using your textbook before searching on the internet. Before you start this sort of task, think carefully about how to structure your study. You might want to have a look at the tips and hints on essay writing (see page 137) before continuing.

Case study

## Is sustainability possible?

As part of his A-level studies, a student was given an open-ended research task about the concept of **sustainability**. He had already been studying this theme as part of his A2 studies and was given a 1,000-word research task. This was the question:

*'Sustainability in human activity is sadly an unachievable goal.'*

Discuss this statement and comment on the extent to which you agree with this view.

Before starting this research project, the student had to make sure he understood the question. It is clearly a very broad question because human activity is so varied, not just within each country but also around the world. He decided to write a short plan before undertaking his research. In it he planned to address some research questions:

1 What is sustainability?
2 What do we mean by human activity?
3 Why is it a goal that human activity should become more sustainable?
4 What does the statement mean by 'sadly'?
5 What human activities are/could become currently unsustainable?
6 What human activities are sustainable?

Following the research, he knew he had to decide on the extent to which he agreed with this view.

### Task

1 Write a 1,000-word report that addresses the research task above.

### Key term

Sustainability: the concept suggesting that future generations should inherit a world which is no less depleted of resources or living conditions than those that we enjoy today.

### Tip

Whenever you are writing a report to a word limit there is usually a tolerance of 10 per cent. This means that you can exceed your recommended word limit by 10 per cent without penalty. Most examination boards pay little attention to word limit, unless there is clear evidence that this has been seriously disregarded.

### Tip

Consult 'command words' in Appendix 2 if you need help.

### Summary

■ Always allow plenty of time to complete internet research.
■ Search engines will give you 'information overload', so choose search terms carefully.
■ Avoid using commercial websites as these require some sort of subscription.
■ Always keep a detailed record of the sites you visit.
■ Try to start your research with a plan and some clearly defined objectives.
■ When handing in a research project, always carefully reference your sources and try to stick fairly closely to the word limit.

Would opponents of a wind farm consider this to be a sustainable energy source?

## GCSE AS/A2 How to revise

Revising is a skill in its own right. Like any other skill, it takes time to learn how to revise effectively and you really do have to think about how you learn best. Some people work better in the morning, while others work better in the evening. It is often assumed that you will automatically know how to revise. However, there is no correct way of revising. Instead, you have to find out what works best for you. Let's consider some strategies and techniques for revising.

Without doubt, the hardest part of revising is getting started with the right level of motivation and self-discipline. Unlike homework, which comes in small, manageable tasks, revision can at first appear overwhelming. At the beginning there appears to be so much to do. You have to start early, stick to your plan and revise in sessions of 45 minutes to 1 hour. You need to take plenty of breaks and sleep well when you are revising. Trying to cram with very little sleep usually becomes counter-productive. Also when revising, try to do at least two sessions each school evening. At the weekends, you should aim to do a lot more, but you will still need to rest between sessions.

### In this section you will learn:

1 the importance of revision
2 the importance of motivation
3 some of the different techniques for revising.

### Key points

There are many ways of revising including:
- reading over your notes and course textbook
- making your own set of revision notes
- using the internet
- working in pairs
- working with private tuition
- voice recording
- watching TV programmes about geography
- using kinaesthetic learning techniques
- using past questions.

### Case study

A student was preparing for his GCSEs and decided to create his own simple revision plan. He started revising at Easter for his summer series of examinations. He needed to cover all his GCSE subjects in his revision programme, though he knew some subjects needed more revision than others.

He wrote his calendar for a period of 10 weeks prior to the beginning of his examinations.

| Day | Session 1 | Session 2 |
|---|---|---|
| Monday 18 April | Geography – rivers: upland processes and landforms | History – Hitler's rise to power |
| Tuesday 19 April | French – verbs | English literature – Of Mice and Men + past questions |
| Wednesday 20 April | Mathematics – trigonometry | English literature – The Tempest |
| Thursday 21 April | History – causes of WW2 | Geography – lowland river landforms and processes |
| Friday 22 April | Biology – the respiratory system | Maths – equations |
| Saturday 23 April | French – speaking | Economics – supply and demand |
| Sunday 24 April | Chemistry – the periodic table | Biology – the reproductive system |

Your geography case studies are a really useful way of covering for the fact that you might not have done enough revision. In longer GCSE questions and A-level questions, you will be expected to use case studies to support your answers. If your specification has a recommended case study, it is a good idea to stick to that particular case study, especially if you are running out of time with your revision. Your course textbook will have a series of case studies, and by sticking to these you will be making it clear to the examiner that you are familiar with your course textbook. However, it is equally worthwhile developing case studies of your own. This just takes a little longer to research. See page 131 for more on researching your own case studies.

## Techniques for revision

There are a few different ways to approach your revision. You will quickly learn what works best for you. Make sure that you vary the way you revise to avoid getting bored otherwise the information won't sink in.

### Tasks

1  Obtain a copy of your specification from your teacher. If you know the specification name, you should be able to get this from the internet.
2  Choose a topic to revise.
3  For GCSE go to www.bbc.co.uk/schools/gcsebitesize/geography and select that topic or for A-level study, select a website from Appendix 1.
4  Make sure that you only cover the elements in your specification.
5  Use your textbook to make revision notes on the same topic.
6  Decide which was the best technique mentioned in the key points on page 135.
7  Practise with some of the other revision techniques.
8  Pick three or four techniques and stick to those as the main ones for your revision programme.

Working in pairs or small groups can be an effective way to revise

> **Tip**
> Don't get too stressed if you have not managed to revise everything before your examinations start. Just make sure that you are ready for the first few. You'll be surprised at how much time you have in between examinations to do last-minute revision.

> **Tip**
> See useful websites in Appendix 1 – these can really make a difference, whatever the level of your study.

> **Tip**
> See Appendix 2 for a detailed explanation of command words; understanding these is a key to success at both GCSE and A-level.

### Summary
- Start your revision early, but not too early.
- Get plenty of rest between revision sessions.
- Have a revision plan.
- Learn your preferred learning style and use it to your advantage.
- Remember that you can also do plenty of revision between your examinations.

# A2 Writing essays

The essay is one area where you cannot afford to be complacent at A2 level. Many specifications use the essay as one way of demonstrating **synopticity**. To write a good essay you need to have a plan and a structure, and make sure that you are answering the question set.

## The plan

A very brief plan is a good idea. In the exam putting a line through your plan tells the examiner to disregard it. In reality though, the examiner is likely to pay little attention to your plan even if you do not put a line through it.

## The features of a well-written essay

**Breadth** refers to the number of relevant and valid points you make throughout your answer. Making many connected points throughout your answer will gain you marks.

**Depth** is concerned with the way each one of your points is explored. Explaining things in detail is important, but you should not get stuck on one point for too long.

**Structure** refers to the way in which you lay out your essay. Try to aim for the following structure as a very rough guide:

| Stage in the essay | % of the overall response | Likely content |
|---|---|---|
| Introduction | 5% | Sets out the direction of the essay. May make some evaluative points or even come to a view. May refer to case studies. May define some of the key terms. Likely to be one or two paragraphs only. |
| Main | 90% | Main body of the response should be split up into several paragraphs. Contains breadth, depth, support and synoptic content. |
| Summary/ conclusion | 5% | Makes a clear, explicit, concise summary or conclusion in relation to the question set. May point to further issues outside the scope of the essay. |

**Support** relates to the way in which you use case studies, place names, specific dates and data. It is always a good tip to have a 'bank' of relevant case studies for each topic you are studying.

**Focus** on the question. Too many students fall into the trap of telling the examiner everything they know about a given topic without clearly answering the question set. One of the clear differentiators between a satisfactory answer and a very good answer is the way in which the candidate remains focused upon the question throughout. This may be explicit, with direct reference to the question in the answer. It may be done implicitly through the theme and content of the points being made.

### In this section you will learn:

1 why essay writing is so important to the assessment process at A-level

2 hints and tips on how to write a good essay.

### Key term

**Synopticity**: your ability to show you understand links between different aspects of geography, for example linking coastal management with tourism or considering global issues, such as global warming. Showing evidence of synopticity is one way of accessing the highest marks in your examination at A2.

The key to using synopticity effectively is relevance. It is not a good idea simply to 'drop in' some unconnected geographical issues that has no bearing on what you are currently writing about. Essay titles will be written in such a way by examiners that it is hard not be synoptic in your answer.

Here are more examples:

- Bringing in a range of scales to your writing.
- Considering contrasting case studies (e.g. European versus African/Asian case studies).
- Incorporating an environmental issue such as acid rain into your human studies, for example transport systems.
- Considering pertinent political issues such as the credit crunch of 2008.

## Case study

### Essay writing in practice

A student was given the following question in preparation for her A-level examination:

*'The process of globalisation has created more costs than benefits.'* To what extent do you agree with this statement?

The first thing she addressed was the command word. Here, it is found in the expression 'To what extent …'. This command is not one word; rather, the candidate is being asked to weigh up the evidence and come to a conclusion. The student decided that the question was asking her to:

- define globalisation
- present the evidence of costs and benefits of the process of globalisation
- evaluate the evidence
- conclude by giving the extent to which she felt globalisation was a positive or negative process.

She used this as the basis for her essay plan.

### Tasks

1  If the globalisation theme above is part of your A-level studies, go on and complete this essay in full. Alternatively, choose an essay from your past papers or specimen papers.

2  Write a plan using the model above.

3  Write an essay in the same format as that suggested on page 137.

4  Obtain a copy of your A-level specification.

5  Find out which A2 examinations require you to show evidence of synopticity.

6  In your revision, make sure that you have identified plenty of potential synoptic content.

7  Read past paper mark schemes or specimen mark schemes from your examination board.

8  Explain how synoptic content is examined in your A-level.

9  In the mark schemes, identify features of the highest mark band answers.

10  Make notes on what you need to do to score the highest marks in the A2 examinations.

### Tip

In order to score top marks in your A2 examination you are almost certain to have to write synoptically. Check your specification carefully to find out which units have a synoptic element. Synopticity may not only be required in an essay question; issue evaluation papers are another example of where you might be required to demonstrate synoptic linkages.

### Summary

- Essay writing is a difficult skill to acquire, so work hard at it.
- Remember breadth, depth, support and structure, and focus on the question when writing an essay.
- Do not spend too long on the plan in the examination.
- Synoptic content should come naturally when essay writing.
- If you want to get an A* at A-level, you need to score very high marks at A2. Scoring high marks at AS is not enough.

**Practising with sample questions**

Once you have completed your examination paper, it will be sent off to the examiner. You then have to wait until your mark comes back. During this process, the examiner will be marking your work against an agreed mark scheme. This same mark scheme will be applied to all the students in the country sitting this paper. There are two ways in which your work will be marked, tick marking and levels marking.

*Tick marking*

A tick (or single mark) is given to each valid point you make. This sort of marking usually only applies to questions worth up to four marks, though this varies depending upon the examination board. For these questions, you should try to achieve the same number of points required by the question. For instance if the question is worth 1 mark, you will probably only have to give a single sentence or one-word answer. Look at this sample GCSE question:

> A student was faced with the question in Task 1 in an examination. It was about countries that are trying to increase their birth rates and began with this quote:
>
> *'Britain's fertility rate – a mere 1.66 per woman – seems positively puny in the face of the news that French women are now producing an average of two children each … France, for instance, has always been famously pro-natal, and today a mixture of cheap and plentiful childcare, generous maternity leave on almost full pay (for 40 weeks for the third child), and other grants, allowances and tax benefits mean that French mothers (who make up a large part of the workforce) are far less penalised for having babies than their British counterparts. So yes, Britain does need more babies, but we'll only get them if we start properly supporting those who make them.'*

Case study

**Task**

1 Describe how France has successfully increased its birth rate. (4 marks)

*Levels marking*

Levels marking is more difficult to understand. This type of marking is usually applied to all questions that are worth more than 4 marks, depending upon the specification. In levels marking it is quality, complexity and sophistication of work that earns credit. It is not about making specific points as with tick marking.

> A group of students were given practice questions as part of the revision for their A-level examination. The teacher expected them to write an answer and then swap papers with their partner for marking. The teacher gave the students the mark scheme as well as the questions. The students were asked not to read the mark scheme until after they had completed the question.

Case study

**In this section you will learn:**

1 how to assess your own work

2 how to assess the work of others

3 how to make sense of a mark scheme.

**Tip**

For this sort of question, you should be trying to make four points for four marks.

**Tip**

The quality of written communication (QWC) is used in different ways according to different specifications. This is concerned with how you use the English language in your writing. It is not specifically concerned with your geographical knowledge. Writing in a clear, concise way is a simple way of meeting the requirements of QWC.

## Tasks

2 Describe the patterns forecast in Resource A. (7 marks)

3 Use Resource B to identify and explain the human factors that lead to enhanced global warming. (10 marks)

4 Referring to specific examples, describe some of the likely impacts of global warming and explain why global warming presents such a great international challenge. (8 marks)

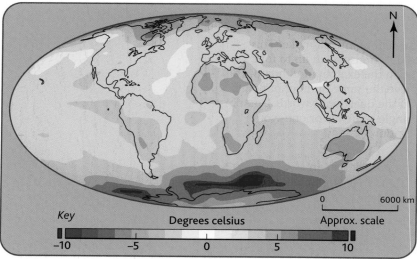

Resource A: Surface air temperature increase, 1960 to 2060

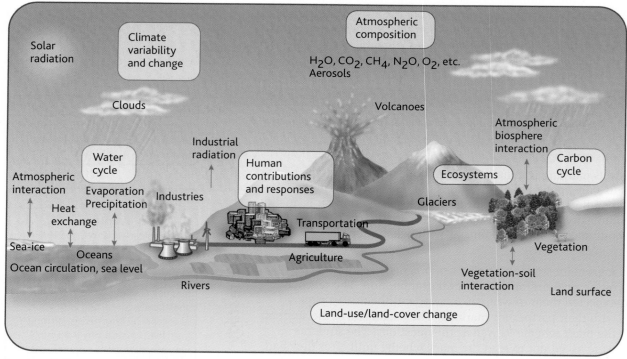

Resource B: Factors contributing to climate change

# A sample A-level fieldwork question

## Mark scheme: notes for answers

### Aim(s)

The aim(s) should be clearly linked to achievable outcomes and be realistic in terms of geographical scale and timescale. There should also be clear evidence of location. The aim(s) should have clear geographical value and significance.

**Example: soils enquiry**
*An investigation into the factors affecting the formation of soils on a hillside in Llanymynech, Wales.*

### Theory

The nature of the supporting theory obviously depends upon the type of enquiry. Expect references to key A-level textbooks, as this is entirely legitimate. Beware of outdated theory that is no longer relevant and watch out especially for theory that is clearly not linked to preceding aim(s). Theory should explore the fundamental processes at work in the chosen enquiry.

**Example: soils enquiry**
Expect theory to examine the influence of relief, rock type and climate, with specific examples of soil types such as podsols, gleys and brown earths. Findings of previous studies might also feature.

### Risk assessment

There should be more than one risk identified. You should go beyond listing and outline the specific nature of the risks within the context of the location.

**Example: soils enquiry**
'I was worried about slipping over on the hillside ...' is a risk, but is clearly very basic and could be anywhere. It is also likely to form part of a list.

'I/We planned our enquiry in a disused quarry, which is now a nature reserve. This presented the problems of loose stones, dangerous under foot. The specific risk was ankle twists and falling on the uneven ground.'

## Minimising risks

You should make specific reference to best practice in health and safety matters. A common-sense response to risk is acceptable, as long as it is clearly linked to the preceding risks.

**Example: soils enquiry**
'I/We identified potential risks before even visiting the site by researching other studies. At the site, I wore sturdy footwear and only conducted my primary data collection in good light. We worked in pairs just in case one of us did get injured. We took a mobile telephone and made sure to inform people of our whereabouts.'

## Mark scheme: the levels

### Level 1

A very basic grasp of the aim(s) and link to theory, which gives the distinct impression that the candidate has not understood the importance or even meaning of risk assessment. Expect vague reference to theory and listing of risks. Risks and responses to risks could be anywhere and not specifically linked to the study site. There may be only one risk considered. One or more of the four aspects is likely to be missing or barely explored. Max. Level 1 if two elements of the answer are missing. Limited QWC.

**(1–4 marks)**

### Level 2

Clear, appropriate aim(s) that are clearly appropriate to a small-scale, time-specific enquiry. Theory may be limited and lacking in depth, especially in relation to the specific aim(s). Risks should be outlined, but perhaps at a superficial and rather simplistic level. Responses to risks may be straight reversals. Max. level 2 if one element of the answer is missing. Satisfactory QWC.

**(5–8 marks)**

### Level 3

For Level 3, all elements must be present. Clear, appropriate aim(s) that are clearly appropriate to a small-scale and/or time-specific enquiry. Theory should be more detailed, but may still lack depth. It should largely relate to the specific aim(s) though. Risks should be outlined and give a real sense that candidate experienced the process of risk assessment. Responses to risks are thoughtful, common sense and even imaginative, though could still legitimately be reversals. Effective QWC.

**(9–12 marks)**

# Appendices

## 1  Useful websites

The table of useful websites below will help you get started when you embark on internet research. There are far too many websites available to the geographer to name here, but this selection provides a basis for your research and many sites listed below will link to other useful websites. Just remember to keep a list of your websites, whether by adding to favourites (or bookmarking your sites) or keeping a handwritten list.

Also included is a set of teacher-specific websites. You can use these sites if you really get stuck on a particular theory or concept. You may also find these sites useful for last-minute revision or if you have been off school for a while and need to catch up on missed work.

| Website | Audience | Area of study | Detail | Key Stage |
|---|---|---|---|---|
| www.games4geog.com | Students | Revision | Fun games and activities to help you revise. | 3 and 4 |
| www.geographyteachingtoday.org.uk | Teachers | Catch-up/revision | A really up-to-date website with lots of new geography. | 3 and 4 |
| www.geography-site.co.uk | Students and teachers | A general geography website for research/catch-up and revision | A very wide selection of themes. | 4 and 5 |
| www.radicalgeography.co.uk/index.html | Teachers | General revision and fun activities | A wide selection of themes. | 3 and 4 |
| www.bucksgfl.org.uk | Teachers | Lots of resources and ideas | Mainly useful for teachers, but could be useful for revision. | 4 and 5 |
| www.gatm.org.uk | Students and teachers | Videos on a range of themes | Educational video clips and PowerPoint presentations on a range of themes. | 3 and 4 |
| www.geographyinthenews.rgs.org/news | Students and teachers | Research using current affairs | Lots of recent news articles on geography-related themes. | 4 and 5 |
| www.geographyalltheway.com | Students and teachers | Catch-up revision | Online resources on a range of themes. | 4 and 5 |
| www.sln.org.uk/geography | Students and teachers | A discussion forum for teachers | Mainly for teachers to share ideas, but great for research or revision. | 5 |
| www.geography.learnontheinternet.co.uk | Students | Catch–up/revision/research | Learning on a range of geography-related topics. | 3 and 4 |
| www.ons.gov.uk/census/index.html | Students and teachers | Census data | A vast database of statistics. | 4 and 5 |
| www.demographia.com | | Population | Lots of population statistics. | 4 and 5 |
| www.migrationinformation.org | Students and teachers | Migration | Information on migration. | 4 and 5 |

| Website | Audience | Area of study | Detail | Key Stage |
|---|---|---|---|---|
| www.prb.org | Teachers and students | Country data | A range of statistical information about different countries. | 5 |
| https://www.cia.gov/library/publications/the-world-factbook/index.html | Students and teachers | Country data | A range of statistical information about different countries. | 4 and 5 |
| www.climateprediction.net | Students and teachers | Climate change | Packed full of resources on climate change. | 4 and 5 |
| www.greenpeace.org/international | Students and teachers | Environmental issues | Full of resources on general environmental issues. | 4 and 5 |
| www.oxfam.org.uk | Students and teachers | Tackling poverty and inequality | Resources and ideas for tackling poverty and inequality. | 4 and 5 |
| www.metoffice.gov.uk/index.html | Students and teachers | Weather and climate | Resources and data on weather and climate around the world. | 4 and 5 |
| www.multimap.com | Students and teachers | Electronic maps at various scales around the UK | Free maps to add to your fieldwork report. | 4 and 5 |
| www.yell.com/ukmaps/home.html | Students and teachers | Electronic maps at various scales around the UK | Clickable map of the UK. Zoom in for a bird's eye view which gives oblique aerial photographs. | 3, 4 and 5 |
| www.upmystreet.com | Students | Simplified local area statistics | A good starting point for gathering information on local areas. | 3 and 4 |
| www.swisseduc.ch/glaciers/earth_icy_planet/glaciers02-en.html | Students | Glacier case studies | A great starting place for researching ice. | 5 |
| www.thewaterpage.com | Students | Water resources | A good starting point for investigating water. | 4 |
| www.un.org/english | Students and teachers | United Nations website- Research | A huge database with information on most aspects of geography. | 5 |
| www.globaleye.org.uk | Students | Research | A series of magazines in 'student-friendly' language. | 3 and 4 |
| www.unwto.org/index.php | Students and teachers | Tourism | Case studies, policies and statistics on world tourism. | 4 and 5 |
| www.fairtrade.org.uk | Students and teachers | Fair trade | A range of teaching and learning resources around the issue of global trade. | 4 and 5 |
| www.environment-agency.gov.uk | Students and teachers | Environment | A range of topics such as air pollution, rivers and farming. Also includes local area information. | 4 and 5 |

# 2 Understanding command words

In the examination itself, you will be under a lot of pressure to complete a paper within a set time limit. It is very easy to misunderstand what the examiner is asking you to do. You have to remain calm and focus clearly upon the question and the specific command words within the question. Here are some of the main command words you will face at GCSE and A-level.

The command word is the specific instruction that you are expected to follow. Get this right and you will score well. Get this wrong and you will score nothing.

| Command word | GCSE, A-level or both | Meaning |
|---|---|---|
| Analyse | A-level | Work out the meaning, usually by breaking something complex down into its components. It can involve manipulating data. |
| Argue | A-level | Present a case. You may need to balance conflicting views. |
| Assess | A-level | Weigh up the evidence and come to a conclusion. |
| Comment | A-level | Offer your opinion on something. |
| Compare | GCSE and A-level | State or compare: discuss the similarities and differences. |
| Complete | GCSE and A-level | Finish off something already started (usually a graph). |
| Contrast | A-level | State or discuss the differences/depending on context of question. |
| Define | GCSE and A-level | State the generally agreed meaning of something – often key geographical terms. |
| Describe | GCSE and A-level | State the main characteristics. Describing questions are often related to a visual stimulus of some kind. |
| Discuss | A-level | Present a range of points connected to a topic or theme. Like 'argue', you may need to present different sides of an argument or conflict. |
| Estimate | A-level | Give a reasoned approximation. |
| Evaluate | GCSE and A-level | Make a value judgement from available information. |
| Examine | A-level | Look at in detail. |
| Explain | GCSE and A-level | Give reasons. Write about why something happens. |
| Identify | GCSE and A-level | Point out key features or characteristics. |
| Interpret | A-level | Make sense of or simplify a more complicated piece of information or situation. |
| Investigate | GCSE and A-level | Look into something in detail. For example, investigate a relationship. |
| Justify | A-level | Explain a position or decision. Say why something should or should not happen. |
| Outline | GCSE and A-level | Set out the basic features or characteristics. Less detailed than 'describe'. |
| State | GCSE and A-level | Identify the correct answer. |
| Summarise | GCSE and A-level | Simplify a more complex situation or piece of information by removing the detail and sticking to the main points. |

# 3 Using OS Landranger maps

## Key to the 1:50 000 Landranger map series

### Communications

## ROADS AND PATHS   Not necessarily rights of way

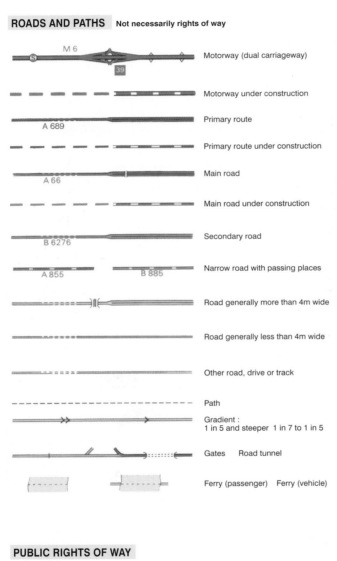

M 6 — Motorway (dual carriageway)

Motorway under construction

A 689 — Primary route

Primary route under construction

A 66 — Main road

Main road under construction

B 6276 — Secondary road

A 855   B 885 — Narrow road with passing places

Road generally more than 4m wide

Road generally less than 4m wide

Other road, drive or track

Path

Gradient :
1 in 5 and steeper  1 in 7 to 1 in 5

Gates      Road tunnel

Ferry (passenger)   Ferry (vehicle)

## PUBLIC RIGHTS OF WAY

.................... Footpath

— — — — — Road used as a public path

—·—·—·—·— Bridleway

-+-+-+-+-+- Byway open to all traffic

## OTHER PUBLIC ACCESS

• • • Other route with public access

◆ National Trail, Long Distance
Route, selected Recreational
Paths

The symbols show the defined route so far
as the scale of mapping will allow. Rights of
way are not shown on maps of Scotland.

• • • National/Regional Cycle Network

8  4  National/Regional Cycle Network number

– – – Surfaced cycle route

The representation on this map of any other
road, track or path is no evidence
of the existence of a right of way.

DANGER AREA  Firing and Test Ranges in the area. Danger! Observe
warning notices.

## RAILWAYS

Track multiple or single

Track under construction

Light rapid transit system, narrow gauge or tramway

Bridges, Footbridge

Tunnel

Station, (a) principal

Siding

Light rapid transit system station

LC  Level crossing

Viaduct

### Tourist Information

Ⅹ Camp Site

⌂ Caravan Site

✿ Garden

⚑ Golf Course or links

ℹ ℹ Information Centre,
all year / seasonal

🦆 Nature reserve

P  P&R
   P&R  Parking

✕ Picnic Site

PC  Public Convenience
(in rural areas)

Selected places of
tourist interest

☎ ☎ Telephone, public/
motoring organisation

☀ Viewpoint

V  Visitor centre

! Walks / Trails

▲ Youth Hostel

## General Information

### LAND FEATURES

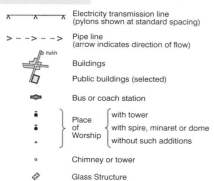

| | |
|---|---|
| Electricity transmission line (pylons shown at standard spacing) | |
| Pipe line (arrow indicates direction of flow) | |
| Buildings | |
| Public buildings (selected) | |
| Bus or coach station | |

Place of Worship
- with tower
- with spire, minaret or dome
- without such additions

Chimney or tower

Glass Structure

Heliport

Triangulation pillar

Mast

Windpump/wind generator

Windmill with or without sails

Graticule intersection at 5' intervals

Cutting, embankment

Quarry

Spoil heap, refuse tip or dump

Coniferous wood

Non-coniferous wood

Mixed wood

Orchard/Park or ornamental grounds

Forestry Commission land

National Trust – always open

National Trust – limited access, observe local signs

National Trust for Scotland

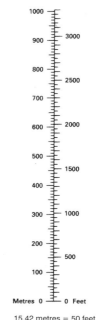

Metres 0    0 Feet

15.42 metres = 50 feet

### BOUNDARIES

| | |
|---|---|
| National | |
| District | |
| County, Unitary Authority, Metropolitan District or London Borough | |
| National Park | |

### ARCHAEOLOGICAL & HISTORICAL INFORMATION

| | | | |
|---|---|---|---|
| + | Site of monument | ☆ ···· | Visible earthwork |
| · o | Stone monument | VILLA | Roman |
| ⚔ | Battlefield (with date) | Castle | Non-Roman |

Information provided by English Heritage for England and the Royal Commissions on the Ancient and Historical Monuments for Scotland and Wales

### WATER FEATURES

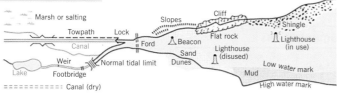

Contour values in lakes are in metres

### ROCK FEATURES

### ABBREVIATIONS

| | |
|---|---|
| CH | Clubhouse |
| MS | Milestone |
| PC | Public convenience (in rural areas) |
| TH | Town Hall, Guildhall or equivalent |
| CG | Coastguard |
| P | Post office |
| MP | Milepost |
| PH | Public house |

### HEIGHTS

—50—    Contours are at 10 metres vertical intervals

·144    Heights are to the nearest metre above mean sea level

Where two heights are shown, the first height is to the base of the triangulation pillar and the second (in brackets) to the highest natural point of the hill.

# 4 Using OS Explorer maps

## Key to the 1:25 000 Explorer map series

### Selected Tourist and Leisure Information

 Parking / Park & Ride

 Telephone, public / motoring organisation

 Camp site / caravan site

 Information centre, all year / seasonal

 Visitor centre

 Forestry Commission Visitor Centre

 Recreation / leisure / sports centre

PC Public convenience

 Picnic site

Golf course or links

 Viewpoint

 Cycle trail

 Country park

 Museum

 Nature reserve

English Heritage

National Trust

Other tourist feature

### General Information

**BOUNDARIES** Administrative boundaries as notified to October 1997

— · — + — · National

— · — · — County

— — — Unitary Authority (UA), Metropolitan District (Met Dist), London Borough (LB) or District

· · · · · · Civil Parish (CP) or Community (C)

— — — Constituency (Const) or Electoral Region (ER)

▬▬ ▬▬ National Park boundary

**VEGETATION** Limits of vegetation are defined by positioning of symbols

Coniferous trees

Non-coniferous trees

Coppice

Orchard

Scrub

Bracken, heath or rough grassland

Marsh, reeds or saltings

**GENERAL FEATURES**

Current or former with tower
place of worship with spire, minaret or dome

+ Place of worship

Building; Important building

Glasshouse

▲ Youth hostel

■ Bunkhouse/camping barn/ other hostel (selected areas only)

Bus or coach station

Lighthouse; disused lighthouse; beacon

Triangulation pillar; mast

Windmill, with or without sails

Wind pump; wind generator

pylon pole Electricity transmission line

Slopes

Gravel pit

Other pit or quarry

Sand pit

Landfill site or slag heap

**HEIGHTS AND NATURAL FEATURES**

Surface heights are to the nearest metre above mean sea level. Where two heights are are shown, the first height is to the base of the triangulation pillar and the second (in brackets) to the highest natural point of the hill.

52 • Ground survey height
284 • Air survey height

Vertical face/cliff

Loose rocks   Boulders   Outcrop   Scree

Contours are at 5 metres vertical intervals

Water

Mud

Sand; sand & shingle

### Communications

**ROADS AND PATHS** Not necessarily rights of way

M 1 or A 6(M) Motorway ⑤ Service area ⑦ Junction number

A 35 Dual carriageway

A 30 Main road

B 3074 Secondary road

Narrow road with passing places

Road under construction

Road generally more than 4 m wide

Road generally less than 4 m wide

Other road, drive or track, fenced and unfenced

Gradient: steeper than 20% (1 in 5); 14% (1 in 7) to 20% (1 in 5)

 Ferry; Ferry P – passenger only

Path

**RAILWAYS**

Multiple track   Standard gauge
Single track

Narrow gauge
Light rapid transit system (LRTS), station

Road over; road under; level crossing

Cutting; tunnel; embankment

Station, open to passengers; siding

**PUBLIC RIGHTS OF WAY**

Footpath
Bridleway
+++++ Byway open to all traffic
Road used as a public path

**OTHER PUBLIC ACCESS**

Other routes with public access (not normally shown in urban areas)

National Trail/Long Distance Route; Recreational Route

Permitted footpath*

Permitted bridleway*
* Footpaths and bridleways along which landowners have permitted public use but are not public rights of way. The agreement may be withdrawn

Traffic-free cycle route

National cycle network route number – traffic-free

National cycle network route number – on road

© Crown Copyright

148

# Explorer map of part of Dartmoor

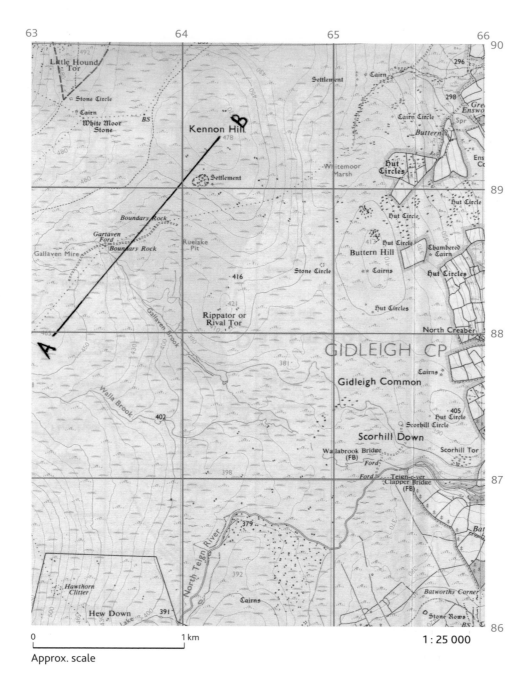

0       1 km

Approx. scale

1 : 25 000

# 5 Critical values tables

## Critical values of chi-squared

| Degrees of freedom | Significance level | |
|---|---|---|
| | 0.05 | 0.01 |
| 1 | 3.84 | 6.64 |
| 2 | 5.99 | 9.21 |
| 3 | 7.82 | 11.34 |
| 4 | 9.49 | 13.28 |
| 5 | 11.08 | 15.09 |
| 6 | 12.59 | 16.81 |
| 7 | 14.07 | 18.48 |
| 8 | 15.51 | 20.09 |
| 9 | 16.92 | 21.67 |
| 10 | 18.31 | 23.21 |
| 11 | 19.68 | 24.72 |
| 12 | 21.03 | 26.22 |
| 13 | 22.36 | 27.69 |
| 14 | 23.68 | 29.14 |
| 15 | 25.00 | 30.58 |
| 16 | 26.30 | 32.00 |
| 17 | 27.59 | 33.41 |
| 18 | 28.87 | 34.80 |
| 19 | 30.14 | 36.19 |
| 20 | 37.57 | 37.57 |
| 21 | 32.67 | 38.93 |
| 22 | 33.92 | 40.29 |
| 23 | 35.18 | 41.64 |
| 24 | 36.43 | 42.98 |
| 25 | 37.65 | 44.31 |
| 26 | 35.88 | 45.64 |
| 27 | 40.11 | 46.96 |
| 28 | 41.34 | 48.28 |
| 29 | 42.56 | 45.59 |
| 30 | 43.77 | 50.89 |
| 40 | 55.76 | 63.69 |
| 50 | 67.51 | 76.15 |
| 60 | 79.08 | 88.38 |
| 70 | 90.53 | 100.43 |
| 80 | 101.88 | 112.33 |
| 90 | 113.15 | 124.12 |
| 100 | 124.34 | 135.81 |

To calculate degrees of freedom where there are A rows and B columns respectively, use DF = (A – 1) × (B – 1). If there is only one row then are (B – 1) degrees of freedom.

Reject $H_0$ if the calculated value of chi-squared is greater than the critical value at the chosen significance level.

# Critical values of Spearman Rank Correlation Coefficient

| Degrees of freedom | Significance level | |
|---|---|---|
| | 0.05 | 0.01 |
| 4 | 1.000 | |
| 5 | 0.900 | 1.000 |
| 6 | 0.829 | 0.943 |
| 7 | 0.714 | 0.893 |
| 8 | 0.643 | 0.833 |
| 9 | 0.600 | 0.783 |
| 10 | 0.564 | 0.745 |
| 11 | 0.523 | 0.736 |
| 12 | 0.497 | 0.703 |
| 13 | 0.475 | 0.673 |
| 14 | 0.457 | 0.646 |
| 15 | 0.441 | 0.623 |
| 16 | 0.425 | 0.601 |
| 17 | 0.412 | 0.582 |
| 18 | 0.399 | 0.564 |
| 19 | 0.388 | 0.549 |
| 20 | 0.377 | 0.534 |
| 21 | 0.368 | 0.521 |
| 22 | 0.359 | 0.508 |
| 23 | 0.351 | 0.496 |
| 24 | 0.343 | 0.485 |
| 25 | 0.336 | 0.475 |
| 26 | 0.329 | 0.465 |
| 27 | 0.323 | 0.456 |
| 28 | 0.317 | 0.448 |
| 29 | 0.311 | 0.440 |
| 30 | 0.305 | 0.432 |

Degrees of freedom = number of paired measures in total sample.

Reject $H_0$ if the value exceeds the critical value at the chosen confidence limit.

# Critical values of Pearson Product Moment Correlation Coefficient

| Degrees of freedom | Significance level | |
|---|---|---|
| | 0.05 | 0.01 |
| 1 | 0.9877 | 0.995 |
| 2 | 0.900 | 0.980 |
| 3 | 0.805 | 0.934 |
| 4 | 0.729 | 0.882 |
| 5 | 0.669 | 0.833 |
| 6 | 0.622 | 0.789 |
| 7 | 0.582 | 0.750 |
| 8 | 0.549 | 0.716 |
| 9 | 0.521 | 0.685 |
| 10 | 0.497 | 0.658 |
| 11 | 0.476 | 0.634 |
| 12 | 0.458 | 0.612 |
| 13 | 0.441 | 0.592 |
| 14 | 0.426 | 0.574 |
| 15 | 0.412 | 0.558 |
| 16 | 0.400 | 0.543 |
| 17 | 0.389 | 0.529 |
| 18 | 0.378 | 0.516 |
| 19 | 0.369 | 0.503 |
| 20 | 0.360 | 0.492 |
| 25 | 0.323 | 0.445 |
| 30 | 0.296 | 0.409 |
| 35 | 0.275 | 0.381 |
| 40 | 0.257 | 0.358 |
| 45 | 0.243 | 0.338 |
| 50 | 0.231 | 0.322 |
| 60 | 0.211 | 0.295 |
| 70 | 0.195 | 0.274 |
| 80 | 0.183 | 0.257 |
| 90 | 0.173 | 0.242 |
| 100 | 0.164 | 0.230 |

Degrees of freedom = $N - 1$ where N is the number of paired observations.

Reject $H_0$ if the calculated value is greater than the critical value (in absolute terms) at the chosen significance level.

# Critical values of the Nearest Neighbour Index

| n | Clustered pattern | | Dispersed pattern | |
|---|---|---|---|---|
| | 0.05 | 0.01 | 0.05 | 0.01 |
| 2 | 0.392 | 0.140 | 1.608 | 1.860 |
| 3 | 0.504 | 0.298 | 1.497 | 1.702 |
| 4 | 0.570 | 0.392 | 1.430 | 1.608 |
| 5 | 0.616 | 0.456 | 1.385 | 1.544 |
| 6 | 0.649 | 0.504 | 1.351 | 1.497 |
| 7 | 0.675 | 0.540 | 1.325 | 1.460 |
| 8 | 0.696 | 0.570 | 1.304 | 1.430 |
| 9 | 0.713 | 0.595 | 1.287 | 1.406 |
| 10 | 0.728 | 0.615 | 1.272 | 1.385 |
| 11 | 0.741 | 0.633 | 1.259 | 1.367 |
| 12 | 0.752 | 0.649 | 1.248 | 1.351 |
| 13 | 0.762 | 0.663 | 1.239 | 1.337 |
| 14 | 0.770 | 0.675 | 1.230 | 1.325 |
| 15 | 0.778 | 0.686 | 1.222 | 1.314 |
| 16 | 0.785 | 0.696 | 1.215 | 1.304 |
| 17 | 0.792 | 0.705 | 1.209 | 1.295 |
| 18 | 0.797 | 0.713 | 1.203 | 1.287 |
| 19 | 0.803 | 0.721 | 1.197 | 1.279 |
| 20 | 0.808 | 0.728 | 1.192 | 1.272 |
| 21 | 0.812 | 0.735 | 1.188 | 1.266 |
| 22 | 0.817 | 0.741 | 1.183 | 1.259 |
| 23 | 0.821 | 0.746 | 1.179 | 1.254 |
| 24 | 0.825 | 0.752 | 1.176 | 1.248 |
| 25 | 0.828 | 0.757 | 1.172 | 1.243 |
| 26 | 0.831 | 0.762 | 1.169 | 1.239 |
| 27 | 0.835 | 0.766 | 1.166 | 1.234 |
| 28 | 0.838 | 0.770 | 1.163 | 1.230 |
| 29 | 0.840 | 0.774 | 1.160 | 1.226 |
| 30 | 0.843 | 0.778 | 1.157 | 1.222 |
| 31 | 0.846 | 0.782 | 1.155 | 1.219 |
| 32 | 0.848 | 0.785 | 1.152 | 1.215 |
| 33 | 0.850 | 0.788 | 1.150 | 1.212 |
| 34 | 0.853 | 0.791 | 1.148 | 1.209 |
| 35 | 0.855 | 0.794 | 1.145 | 1.206 |
| 36 | 0.857 | 0.797 | 1.143 | 1.203 |
| 37 | 0.859 | 0.800 | 1.141 | 1.200 |
| 38 | 0.861 | 0.803 | 1.140 | 1.197 |

| | | | | |
|---|---|---|---|---|
| 39 | 0.862 | 0.805 | 1.138 | 1.195 |
| 40 | 0.864 | 0.808 | 1.136 | 1.192 |
| 41 | 0.866 | 0.810 | 1.134 | 1.190 |
| 42 | 0.867 | 0.812 | 1.133 | 1.188 |
| 43 | 0.869 | 0.815 | 1.131 | 1.186 |
| 44 | 0.870 | 0.817 | 1.130 | 1.183 |
| 45 | 0.872 | 0.819 | 1.128 | 1.181 |
| 50 | 0.878 | 0.828 | 1.122 | 1.172 |
| 60 | 0.889 | 0.843 | 1.111 | 1.157 |
| 70 | 0.897 | 0.855 | 1.103 | 1.145 |
| 80 | 0.904 | 0.864 | 1.096 | 1.136 |
| 90 | 0.909 | 0.872 | 1.091 | 1.128 |
| 100 | 0.914 | 0.378 | 1.086 | 1.122 |

$n$ = number of points in the survey.

To test for clustering: reject $H_0$ if the calculated value of $R$ is less than the critical value at the chosen significance level.

To test for dispersion: reject $H_0$ if the calculated value of $R$ is greater than the critical value at the chosen significance level.

# Critical values of student's *t*-test

| Degrees of freedom | Significance level | |
|---|---|---|
| | 0.05 | 0.01 |
| 1 | 6.31 | 63.66 |
| 2 | 2.92 | 9.93 |
| 3 | 2.35 | 5.84 |
| 4 | 2.13 | 4.60 |
| 5 | 2.00 | 4.03 |
| 6 | 1.94 | 3.71 |
| 7 | 1.89 | 3.50 |
| 8 | 1.86 | 3.36 |
| 9 | 1.83 | 3.25 |
| 10 | 1.81 | 3.17 |
| 11 | 1.80 | 3.11 |
| 12 | 1.78 | 3.06 |
| 13 | 1.77 | 3.01 |
| 14 | 1.76 | 2.98 |
| 15 | 1.75 | 2.95 |
| 16 | 1.75 | 2.92 |
| 17 | 1.74 | 2.90 |
| 18 | 1.73 | 2.88 |
| 19 | 1.73 | 2.86 |
| 20 | 1.73 | 2.85 |
| 21 | 1.72 | 2.83 |
| 22 | 1.72 | 2.82 |
| 23 | 1.71 | 2.81 |
| 24 | 1.71 | 2.80 |
| 25 | 1.71 | 2.79 |
| 26 | 1.71 | 2.78 |
| 27 | 1.70 | 2.77 |
| 28 | 1.70 | 2.76 |
| 29 | 1.70 | 2.76 |
| 30 | 1.70 | 2.75 |

To calculate degrees of freedom where the two sample sizes are A and B respectively, $DF = (A - 1) + (B - 1)$.

Reject $H_0$ if the calculated value of $t$ is greater than the critical value at the chosen level of significance.

# The Mann-Whitney $U$-test

Critical values of $U$ at $p = 0.05$

| $n_1$ | 1 | 2 | 3 | 4 | 5 | 6 | 7 | 8 | 9 | 10 | 11 | 12 | 13 | 14 | 15 | 16 | 17 | 18 | 19 | 20 |
|---|---|---|---|---|---|---|---|---|---|---|---|---|---|---|---|---|---|---|---|---|
| 1 | - | - | - | - | - | - | - | - | - | - | - | - | - | - | - | - | - | - | - | - |
| 2 | - | - | - | - | - | - | - | 0 | 0 | 0 | 0 | 1 | 1 | 1 | 1 | 1 | 2 | 2 | 2 | 2 |
| 3 | - | - | - | - | 0 | 1 | 1 | 2 | 2 | 3 | 3 | 4 | 4 | 5 | 5 | 6 | 6 | 7 | 7 | 8 |
| 4 | - | - | - | 0 | 1 | 2 | 3 | 4 | 4 | 5 | 6 | 7 | 8 | 9 | 10 | 11 | 11 | 12 | 13 | 13 |
| 5 | - | - | 0 | 1 | 2 | 3 | 5 | 6 | 7 | 8 | 9 | 11 | 12 | 13 | 14 | 15 | 17 | 18 | 19 | 20 |
| 6 | - | - | 1 | 2 | 3 | 5 | 6 | 8 | 10 | 11 | 13 | 14 | 16 | 17 | 19 | 21 | 22 | 24 | 25 | 27 |
| 7 | - | - | 1 | 3 | 5 | 6 | 8 | 10 | 12 | 14 | 16 | 18 | 20 | 22 | 24 | 26 | 28 | 30 | 32 | 34 |
| 8 | - | 0 | 2 | 4 | 6 | 8 | 10 | 13 | 15 | 17 | 19 | 22 | 24 | 26 | 29 | 31 | 34 | 36 | 38 | 41 |
| 9 | - | 0 | 2 | 4 | 7 | 10 | 12 | 15 | 17 | 20 | 23 | 26 | 28 | 31 | 34 | 37 | 39 | 42 | 45 | 48 |
| 10 | - | 0 | 3 | 5 | 8 | 11 | 14 | 17 | 20 | 23 | 26 | 29 | 33 | 36 | 39 | 42 | 45 | 48 | 52 | 55 |
| 11 | - | 0 | 3 | 6 | 9 | 13 | 16 | 19 | 23 | 26 | 30 | 33 | 37 | 40 | 44 | 47 | 51 | 55 | 58 | 62 |
| 12 | - | 1 | 4 | 7 | 11 | 14 | 18 | 22 | 26 | 29 | 33 | 37 | 41 | 45 | 49 | 53 | 57 | 61 | 65 | 69 |
| 13 | - | 1 | 4 | 8 | 12 | 16 | 20 | 24 | 28 | 33 | 37 | 41 | 45 | 50 | 54 | 59 | 63 | 67 | 72 | 76 |
| 14 | - | 1 | 5 | 9 | 13 | 17 | 22 | 26 | 31 | 36 | 40 | 45 | 50 | 55 | 59 | 64 | 67 | 74 | 78 | 83 |
| 15 | - | 1 | 5 | 10 | 14 | 19 | 24 | 29 | 34 | 39 | 44 | 49 | 54 | 59 | 64 | 70 | 75 | 80 | 85 | 90 |
| 16 | - | 1 | 6 | 11 | 15 | 21 | 26 | 31 | 37 | 42 | 47 | 53 | 59 | 64 | 70 | 75 | 81 | 86 | 92 | 98 |
| 17 | - | 2 | 6 | 11 | 17 | 22 | 28 | 34 | 39 | 45 | 51 | 57 | 63 | 67 | 75 | 81 | 87 | 93 | 99 | 105 |
| 18 | - | 2 | 7 | 12 | 18 | 24 | 30 | 36 | 42 | 48 | 55 | 61 | 67 | 74 | 80 | 86 | 93 | 99 | 106 | 112 |
| 19 | - | 2 | 7 | 13 | 19 | 25 | 32 | 38 | 45 | 52 | 58 | 65 | 72 | 78 | 85 | 92 | 99 | 106 | 113 | 119 |
| 20 | - | 2 | 8 | 13 | 20 | 27 | 34 | 41 | 48 | 55 | 62 | 69 | 76 | 83 | 90 | 98 | 105 | 112 | 119 | 127 |

Reject $H_0$ if the calculated value of $U$ is equal to or less than the appropriate critical value.

# Random number table

| | | | | | | | | | | | |
|---|---|---|---|---|---|---|---|---|---|---|---|
| | 17 | 42 | 28 | 23 | 17 | 59 | 66 | 38 | 61 | 02 | 10 |
| | 10 | 51 | 55 | 92 | 52 | 74 | 49 | 04 | 49 | 03 | 04 |
| 33 | 53 | 70 | 11 | 54 | 48 | 63 | 50 | 90 | 37 | 21 | 46 |
| 77 | 84 | 87 | 67 | 39 | 95 | 85 | 54 | 97 | 37 | 33 | 41 |
| 11 | 75 | 74 | 90 | 50 | 08 | 91 | 12 | 44 | 82 | 40 | 30 |
| 62 | 45 | 50 | 64 | 54 | 65 | 17 | 89 | 25 | 59 | 44 | 64 |
| 59 | 33 | 23 | 31 | 39 | 84 | 54 | 33 | 20 | 76 | 25 | 50 |
| 04 | 15 | 26 | 89 | 98 | 17 | 52 | 53 | 82 | 62 | 02 | 21 |
| 82 | 34 | 13 | 41 | 03 | 68 | 97 | 81 | 40 | 72 | 61 | 52 |
| 40 | 49 | 27 | 56 | 49 | 79 | 34 | 34 | 32 | 22 | 60 | 53 |
| 91 | 17 | 08 | 72 | 87 | 46 | 75 | 73 | 00 | 11 | 27 | 07 |
| 05 | 20 | 30 | 85 | 22 | 21 | 04 | 67 | 95 | 97 | 98 | 62 |
| 17 | 27 | 31 | 42 | 64 | 71 | 64 | 00 | 26 | 04 | 66 | 91 |
| 03 | 64 | 59 | 07 | 42 | 95 | 81 | 39 | 06 | 41 | 29 | 81 |
| 90 | 32 | 70 | 17 | 72 | 03 | 61 | 66 | 26 | 24 | 71 | 97 |
| 27 | 26 | 08 | 79 | 61 | 03 | 62 | 93 | 23 | 29 | 26 | 04 |
| 50 | 14 | 30 | 85 | 38 | 97 | 56 | 37 | 08 | 12 | 23 | 07 |
| 61 | 05 | 92 | 08 | 29 | 94 | 10 | 96 | 50 | 01 | 33 | 85 |
| 66 | 28 | 02 | 45 | 37 | 89 | | | | | | |

# 6 Graph paper

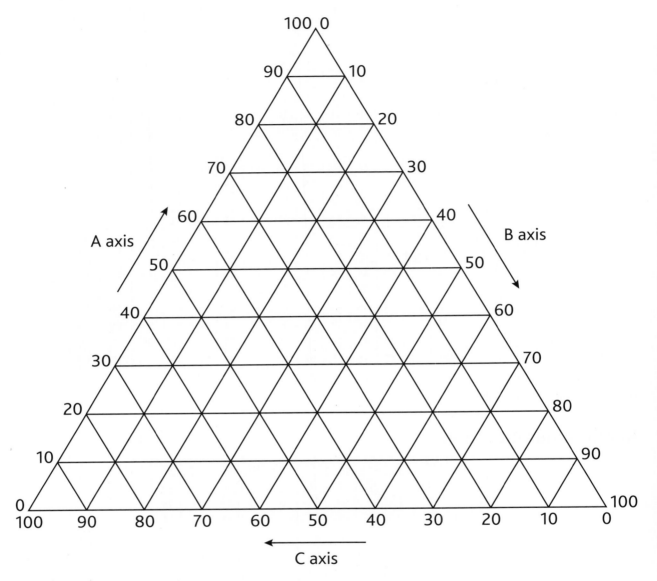

Triangular graph paper

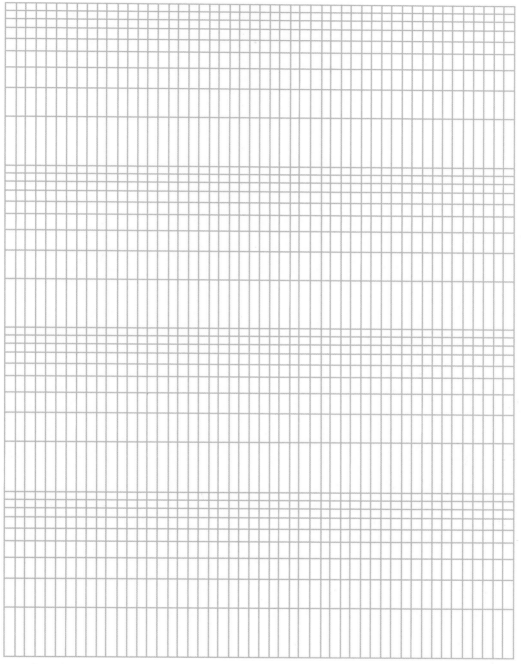

Semi-log graph paper

Log-log graph paper

# Index